YOUR KNOWLEDGE HAS VALUE

Chimere Anabanti

The Whitehead Algorithm for free groups

GRIN Publishing

Bibliographic information published by the German National Library:

The German National Library lists this publication in the National Bibliography; detailed bibliographic data are available on the Internet at http://dnb.dnb.de .

Imprint:

Copyright © 2013 GRIN Verlag GmbH
Print and binding: Books on Demand GmbH, Norderstedt Germany
ISBN: 978-3-656-92267-4

This book at GRIN:

http://www.grin.com/en/e-book/294023/the-whitehead-algorithm-for-free-groups

GRIN - Your knowledge has value

Since its foundation in 1998, GRIN has specialized in publishing academic texts by students, college teachers and other academics as e-book and printed book. The website www.grin.com is an ideal platform for presenting term papers, final papers, scientific essays, dissertations and specialist books.

Visit us on the internet:

http://www.grin.com/

http://www.facebook.com/grincom

http://www.twitter.com/grin_com

The Whitehead Algorithm For Free Groups

Chimere Stanley Anabanti

9th September 2013
Submitted in partial fulfillment of a Masters in Mathematics requirement at the
University of Warwick, United Kingdom.

Abstract

We start with a brief introduction to Free Groups, thereby appreciating Nielsen's approach to the Subgroup theorem. Beautiful results of J. H. C. Whitehead, J. Nielsen, E. S. Rapaport, Higgins and Lyndon, and J. McCool form our building block. We study different automorphisms of a finitely generated free group as well as a finite set of automorphisms which Whitehead used to deduce that if two elements of a finitely generated free group are equivalent under an automorphism of the group, then they are equivalent under such automorphisms. We write program aimed at appreciating Whitehead's theorem, starting with programs for appreciating Whitehead automorphisms to programs for determining whether two elements of a finitely generated free group are equivalent or not. We conclude by classifying all minimal words of lengths $2, 3, 4, 5$ and 6 in F_n (for some $n \in [2, 6]$) up to equivalence.

Contents

1. Introduction to Free Groups

The importance of Free Groups cannot be overempasized. One of its beautiful roles is in Presentations of Groups. We begin this discussion with a description of basic concepts in Free Groups.

1.1 Definition, Existence and Uniqueness Theorem

We start with the Universal property of a free group given as Definition 1.1.1 below.

1.1.1 Definition. A group F is said to be free on a subset $X \subseteq F$ if given any group G, and any map $\theta : X \to G$, $\exists!$ homomorphism $\theta' : F \to G$ (called extension of θ) such that if $i : X \hookrightarrow F$, then $\theta = \theta' i$.

We now define other terms that we shall use from time to time.

1.1.2 Definition. • Let G be a group; and $X \subseteq G$. A **word** over X is an expression of the form $w = x_1^{\delta_1} x_2^{\delta_2} \cdots x_l^{\delta_l}$ (for $x_1, x_2, \cdots, x_l$ elements of X, and $\delta_i \in \{-1, 1\}$ for $1 \leq i \leq l$).

- l is called the **length of the word, w,** and denoted as $l(w)$ or $|w|$.

- The **empty word** is denoted by ϵ or 1, and its length is 0.

- A **subword** of $w = x_1^{\delta_1} x_2^{\delta_2} \cdots x_l^{\delta_l}$ is a word $w_1 = x_d^{\delta_d} \cdots x_m^{\delta_m}$ (where $1 \leq d \leq m \leq l$).

- A word $w = x_1^{\delta_1} x_2^{\delta_2} \cdots x_l^{\delta_l}$ is called **reduced** if it does not admit any subword of the form xx^{-1} or $x^{-1}x$.

- For any set X, we define $A = X \cup X^{-1}$, $A^* = \{$all words over A$\}$, and $F_X = \{$all reduced words over A$\}$.

- A word $w = a_1 a_2 \cdots a_l \in A^*$ is called **cyclically reduced** if it is reduced, and either $l(w) = 0$ or $a_1 \neq a_l^{-1}$.

Next, we state the existence and uniqueness theorem.

1.1.3 Theorem (Existence). For any set X, there is a free group F_X on X.

1.1.4 Theorem (Uniqueness). If F_i is free on X_i $(i = 1, 2)$, then $F_1 \cong F_2 \iff |X_1| = |X_2|$.

Up to isomorphism, a free group can be characterised by its rank.

1.1.5 Definition. Rank of a free group is the cardinality of the generating set of the free group.

1.1.6 Example. • $\mathbb{Z} = \{x^n | n \in \mathbb{Z}\}$ is free of rank 1 (since for any group G and $g \in G$, $\exists!$ homomorphism $\theta' : \mathbb{Z} \to G$ s.t $\theta'(x^n) = g^n \; \forall \, n \in \mathbb{Z}$).

- $id_F = \{\epsilon\}$ is free of rank 0.

In the sequel, we denote a free group on X by F_X. If $|X| = n < \infty$, we represent F_X as F_n.

1.1.7 Remark. There is no relationship between the term "free" used here, and that of a "free action". The term "free" simply tells us that there is no other relation between elements of such group, except any relation induced by a group property such as $xx^{-1} = \epsilon$.

1

1.2 Basic Properties of Free Groups

1.2.1 Proposition. A free group is torsion-free.

Proof.   • Given $\epsilon \neq w \in F$ such that $l(w) = l$ with $w = \beta\alpha\beta^{-1}$ (where $\beta = a_1 a_2 \cdots a_r$, $\alpha = a_{r+1} \cdots a_{l-r}$, and $\beta^{-1} = a_r^{-1} a_{r-1}^{-1} \cdots a_1^{-1}$ as every non-identity element of a free group is conjugate to a cyclically reduced word).

• For $n \geq 2$, $w^n = (\beta\alpha\beta^{-1})^n = \beta\alpha^n\beta^{-1}$ is reduced.

• Now, $l(w^n) = nl(\alpha) + 2l(\beta) > (n-1)l(\alpha) + 2l(\beta) = l(w^{n-1})$. So, for $w \neq \epsilon$, $l(w) < l(w^2) < \cdots$.

• $\therefore \circ(w) = \infty$.

$\square$

1.2.2 Proposition. Any group G is isomorphic to a quotient of a free group.

Proof.   • Let $G = \langle X \rangle$.

• By Definition 1.1.1, the map $\theta : X \to G$ extends uniquely to a homomorphism $\theta' : F_X \to G$.

• θ' is onto (since $G = \langle X \rangle$).

• By first isomorphism theorem therefore, $F_X / Ker(\theta') \cong Im(\theta') = G$.

$\square$

1.2.3 Proposition. A free group of finite rank n cannot be generated by fewer than n elements.

Proof.   • Let F be free on X (where $|X| = n < \infty$).

• Suppose $F = \langle Y \rangle$, and G is a finite group of order m.

• Given a map $\theta : Y \to G$, θ extends to at most one homomorphism $\theta' : F \to G$. So, $|Hom(F, G)| \leq |Map(Y, G)| = m^{|Y|}$.

• As F is free on X, $|Hom(F, G)| = |Map(X, G)| = m^n$.

• Thus, $m^n \leq m^{|Y|}$, and so result follows.

$\square$

1.2.4 Lemma. If F is Free on X, then $F = \langle X \rangle$.

1.2.5 Proposition. Let the group $G = \langle X \rangle$. Then G is free on X if and only if no nonempty reduced word in A^* represents 1_G.

Proof.   • Let $G = \langle X \rangle$.

• ($\Rightarrow$) But F_X is free on X. So by Theorem 1.1.3, $\exists \, \theta : F_X \xrightarrow{\text{isomorphism}} G$.

• As F_X is free on X, no nonempty reduced word in A^* represents 1_{F_X}.

- Thus, by the isomorphism above, no nonempty reduced word in A^* represents 1_G.

- ($\Leftarrow$) Define $\theta : X \to G$ by $\theta(x) = x$.

- By definition of Free groups, θ extends uniquely to a homomorphism $\theta' : F_X \to G$.

- $G = \langle X \rangle \Rightarrow \theta'$ is onto.

- Now, no nonempty reduced word in A^* represents $1_G \Rightarrow \theta'$ is $1 - 1$.

- So, G is free on $\{\theta'(x) : x \in X\}$; and hence G is free on X.

$\square$

1.2.6 Lemma. The number of reduced words of length $l \geq 1$ in a Free group of rank r is $2r(2r-1)^{l-1}$.

Proof.   • Let $w = a_1 a_2 \cdots a_l$ be a reduced word of length l in a Free group of rank r.

- a_1 can be any of the r generators or their inverses.

- $a_i (i \geq 2)$ can be any of the r generators or their inverses, except a_{i-1}^{-1}.

- So, there are $2r$ choices for a_1, and $(2r - 1)$ choices for each a_i, $2 \leq i \leq l$.

- Thus, the number of reduced words of length l in a free group of rank r is $(2r)(2r - 1)^{l-1}$.

$\square$

1.3   Subgroups of Free Groups

We start with an example of a subgroup of a free group as follows:

1.3.1 Lemma. Let F be free on X (where $X = \{x, y\}$). The set of elements of F consisting of reduced words of even length is a subgroup of F generated by the set $\{x^2, xy, xy^{-1}\}$.

Next, we introduce the main result of this section.

1.3.2 Theorem (Nielsen-Schreier). Every subgroup of a free Group is free.

Theorem 1.3.2 is referred to as the Subgroup theorem. There are at least three different proofs of this: two are Algebraic in nature (given by Otto Schreier and Jakob Nielsen), and the third which uses concepts from Algebraic Topology. We discuss only Nielsen's approach owing to the fact that it gives us a beautiful way of appreciating the generators of Automorphisms of Free Groups which we shall consider in the next chapter. Henceforth, our discussions will be restricted to finitely generated free groups for which our interest lies, following [Joh97]. See [LS77] or [Joh97] for a general approach.

Let $F = F_X$ be a free group; and $U = \{u_1, \cdots, u_n\} \subseteq F$.

1.3.3 Definition. An **elementary Nielsen transformation** of U is one of the following three types:

- (A1) delete some u_i where $u_i = \epsilon$;

- (A2) replace some u_i by u_i^{-1};

- (A3) replace some u_i by $u_i u_j$

  (where $1 \leq i \leq n$, and u_k remains the same for $k \neq i$).

1.3.4 Notation. For ease of reference, we denote types (A1), (A2) and (A3) elementary Nielsen transformations by $(\backslash i)$, (i') and (ij) respectively.

1.3.5 Definition. • A **Nielsen transformation** (N-transf for short) is a finite sequence of elementary Nielsen transformations.

- A Nielsen transformation is called **regular** if it does not involve any (A1) operation, and called **singular** if otherwise.

1.3.6 Example. • Let $U_1 = (ab^{-1}, b^{-1}, c, \epsilon)$ and $\tau_1 = (\backslash 4)$. $U_1 \xrightarrow{(\backslash 4)} (ab^{-1}, b^{-1}, c) = U_1 \tau_1 = V_1$.

- Let $U_2 = (ab, a)$ and $\tau_2 = (2')(21)(2')(12)(2')$. Then $U_2 \xrightarrow{(2')} (ab, a^{-1}) \xrightarrow{(21)} (ab, b) \xrightarrow{(2')}$ $(ab, b^{-1}) \xrightarrow{(12)} (a, b^{-1}) \xrightarrow{(2')} (a, b) = U_2 \tau_2 = V_2$.

τ_1 and τ_2 are Nielsen transformations; τ_1 is singular while τ_2 is regular.

1.3.7 Lemma (Basic results).   • (B1) If τ is a Nielsen transformation with m operations of (A1), then $|U\tau| = |U| - m$.

- (B2) If U is carried into V by a Nielsen transformation τ, then $\langle U \rangle = \langle V \rangle$.

- (B3) The regular Nielsen transformations form a group.

1.3.8 Definition. A set $V \subseteq F$ is called **Nielsen reduced** (N-reduced for short) if $\forall\, a, b, c \in V^{\pm 1}$, the following three conditions are satisfied:

- (C1) $a \neq \epsilon$;

- (C2) $ab \neq \epsilon \Rightarrow l(ab) \geq l(a), l(b)$;

- (C3) $ab \neq \epsilon$ and $bc \neq \epsilon \Rightarrow l(abc) > l(a) - l(b) + l(c)$.

1.3.9 Example. • From our previous example, V_2 is N-reduced while V_1 is not N-reduced.

- V_1 is not N-reduced since $l(xyz) = 2 = l(x) - l(y) + l(z)$ for $x = ab^{-1}$, $y = (b^{-1})^{-1}$, $z = c$; thus contradicting (C3) above. Also, (C2) fails.

1.3.10 Proposition. Let $V \subseteq F$ be N-reduced, and $w = v_1 v_2 \cdots v_n$ a reduced word in $V^{\pm 1}$. Then:

- (D1) $l(w) \geq l(v_i)$ $(1 \leq i \leq n)$;

- (D2) $l(w) \geq n$;

- (D3) V is a basis for the subgroup $\langle V \rangle$ of F.

1.3.11 Proposition. If $U = (u_1, u_2, \cdots, u_n)$ is an ordered n-tuple of elements of a free group F_X, then U can be carried by a Nielsen transf into an N-reduced tuple V.

1.3.12 Theorem. Every finitely generated subgroup of a free group is free.

Proof.   • Let F be free with basis X; and $H = \langle U \rangle$ (where $H \leq F$, and $U \subseteq F$ with $|U| < \infty$).

   • By Proposition 1.3.11, U can be carried by a Nielsen transformation into an N-reduced tuple V.

   • By Lemma 1.3.7B2, $\langle U \rangle = \langle V \rangle$; and by Proposition 1.3.10D3, V is a basis for H.

   • Thus, H is free of rank $|V|$.

$\square$

We now give a second proof of Proposition 1.2.3.

1.3.13 Proposition. A free group of finite rank n cannot be generated by fewer than n elements.

Proof.   • Let F be free on X (where $|X| = n < \infty$).

   • Consider a finite subset U of generators of F.

   • By Proposition 1.3.11, U can be carried by a Nielsen transformation into an N-reduced tuple V. By Lemma 1.3.7B2, $\langle U \rangle = \langle V \rangle$; and by Proposition 1.3.10D3, V is a basis for F.

   • As the generating set of a free group is unique, $n = rk(F) = |V|$.

   • Recall: If V is gotten from U by a Nielsen transformation τ, then $|V| = |U|$ or $|V| < |U|$ according as τ being regular or singular.

   • Thus, $|V| \leq |U|$ in both cases, and result follows.

$\square$

1.3.14 Corollary. Let F be a free group of finite rank n, and generated by a set U of n elements. Then U is a basis for F.

MANY THANKS TO NIELSEN!!!

2. Automorphisms of Free Groups

Automorphisms of finitely generated free groups as well as the Whitehead automorphisms will be investigated. Results of J. H. C. Whitehead ([Whi36a] and [Whi36b]), J. Nielsen ([Nie24]), E. S. Rapaport ([Rap58]), Higgins and Lyndon ([HL74]), and J. McCool ([McC74]) will form our building block. We conclude by investigating equivalence of elements under $Aut(F_n)$, and giving a presentation of $Aut(F_n)$.

2.1 Special Automorphisms of F_X

[LS77] referred to the following as the key to the study of Automorphisms of Free groups:

"**If F is free on X, then every automorphism of F carries X onto another basis Y. On the other hand, every 1-1 map from X onto another basis of F defines an automorphism**".

2.1.1 Definition. Let F_X be the free group on X. Any map from X into F_X defines an endomorphism of F_X. We define the **Special Endomorphisms of F_X** as follows:

- For fixed $x \in X$, define $\alpha_x : x \mapsto x^{-1}, y \mapsto y \ \forall \ y \in X - \{x\}$;

- Given $x, y \in X$ (with $x \neq y$), define $\beta_{xy} : x \mapsto xy, y \mapsto y \ \forall \ y \in X - \{x\}$.

2.1.2 Remark. Both α_x and β_{xy} are automorphisms of F_X (since for each special endomorphism, image of X is a basis of F_X). In the sequel, we refer to both α_x and β_{xy} as **Special Automorphisms of F_X**.

2.1.3 Lemma. The Special automorphisms of F_X are elementary regular Nielsen transformations.

2.1.4 Notation.
- The set of all Special automorphisms of F_n will be denoted by ψ_n.

- The subgroup of $Aut(F_X)$ generated by the Special automorphisms will be denoted by N. So, $N = \langle \alpha_x, \beta_{xy} : x, y \in X, x \neq y \rangle$.

2.1.5 Theorem. Let F be free on X; and N as defined above. Then:

- (E1) N is dense in $Aut(F)$: If $u_i \in F$ (for $1 \leq i \leq n$), and $\alpha \in Aut(F)$, then $\exists \ \beta \in N$ such that

$$\alpha(u_i) = \beta(u_i) \ \forall \ i;$$

- (E2) If $|X| = n < \infty$, then $Aut(F_n) = N$.

Proof. See page 23 of [LS77]. □

We now see a necessary and sufficient condition for an automorphism of F to belong to N.

2.1.6 Proposition. An automorphism α of F belongs to N if and only if there exists a basis X of F such that α fixes all but finitely many $x \in X$.

Proof. See page 23 of [LS77]. □

From now onwards, we shall conform to automorphisms of finitely generated Free groups for which the Special automorphisms generate. A natural question is:

2.1.7 Question. Does $Aut(F_n)$ have finite presentation?

Nielsen first answered it by constructing a finite presentation for $Aut(F_n)$ in 1924 as seen in [Nie24]. Fifty years later, McCool in [McC74] gave a simpler finite presentation for $Aut(F_n)$. Many authors also attempted other approaches to deduce the same result. See [AFV08], [Neu32] and [MKS76] for more details. We shall follow McCool's approach in our later discussion. For now, let us appreciate automorphisms of the Abelianization of F_n.

2.2 Automorphisms of $\bar{F}_n$

2.2.1 Definition. The **Abelianization of** F_n is denoted by $\bar{F}_n$, and defined by:

$$\bar{F}_n = F_n/[F_n, F_n],$$

where $[F_n, F_n]$ is the Commutator Subgroup.

2.2.2 Proposition. $Aut(\bar{F}_n) \cong GL(n, \mathbb{Z})$.

2.2.3 Lemma ([LS77]). • (F1) $\bar{F}_n$ is Free Abelian with basis as the image $\bar{X}$ of X in $\bar{F}_n$.

• (F2) The Special automorphisms give rise to generators of $Aut(\bar{F}_n)$.
In particular, $Aut(\bar{F}_n) = \langle \bar{\alpha}_i, \bar{\beta}_{ij} \rangle$, where $\bar{\alpha}_i : \bar{x}_i \mapsto -\bar{x}_i$, and $\bar{\beta}_{ij} : \bar{x}_i \mapsto \bar{x}_i + \bar{x}_j$ for $1 \leq i, j \leq n$ with $i \neq j$.

2.2.4 Definition. We define the map $\delta : Aut(F_n) \rightarrow Aut(\bar{F}_n)$ in the natural way.

2.2.5 Lemma. δ is an epimorphism.

We now pause, and give a Characterization of $Aut(F_2)$ and $Aut(\bar{F}_2)$.

2.2.6 Lemma ([Mes74]). • (G1) $Aut(F_2) = \langle P, X, Y, A, B \mid R \rangle$, with set of relators R given as:
$R = \{X^4 = P^2 = (PX)^2 = 1, (PY)^2 = B, X^2 = Y^3B^{-1}A, A^P = A^X = A^Y = B, B^P = A, B^X = A^{-1}, B^Y = A^{-1}B\}$, where $a^P = a^X = a^Y = b, b^P = a, b^X = a^{-1}, b^Y = a^{-1}b$, and A and B are the inner automorphisms induced by a and b respectively.

• (G2) $Inn(F_2) = \langle A, B \rangle$.

• (G3) $Out(F_2) = Aut(F_2)/Inn(F_2) \cong GL(2, \mathbb{Z})$.

• (G4) $GL(2, \mathbb{Z}) = \langle p, x, y \mid x^4 = p^2 = (px)^2 = (py)^2 = 1, x^2 = y^3 \rangle$, where the generators p, x and y are given as $\left(\begin{smallmatrix} 0 & 1 \\ 1 & 0 \end{smallmatrix}\right), \left(\begin{smallmatrix} 0 & 1 \\ -1 & 0 \end{smallmatrix}\right)$ and $\left(\begin{smallmatrix} 0 & 1 \\ -1 & 1 \end{smallmatrix}\right)$ respectively.

2.2.7 Corollary. Considering F_2, $Ker(\delta) = Inn(F_2)$.

2.2.8 Notation. For $g \in F_2$, we denote the automorphism induced by the conjugation by g as $\hat{g}$, where

$$\hat{g}(x) = g^{-1}xg, \quad x \in F_2.$$

Automorphisms of F_2 and its abelianization have been deeply studied by many authors. We summarize the results of [Mes74], [LS77] and [Bog00] in the following proposition.

2.2.9 Proposition.   • (H0i) Every element of $Aut(F_2)$ can be written in the form:

$$P^l W_1(X,Y) X^{2m} W_2(A,B),$$

where $l, m \in \{0,1\}$, $W_2(A,B)$ is a reduced word, and $W_1(X,Y)$ is a reduced word with X, Y and Y^{-1} as the only terms for which powers of X and Y can occur.

• (H0ii) Similarly, every element of $Aut(\bar{F}_2)$ can be written in the equivalent lower cases.

• (H1i) $Aut(F_2)$ has elements of finite order m only for $m = 1, 2, 3$ and 4.

• (H1ii) $Aut(\bar{F}_2)$ has elements of orders m only for $m = 1, 2, 3, 4$ and 6.

• (H2i) A periodic element of $Aut(F_2)$ is conjugate to exactly one of $P, PX, PXA, X^2, Y^2 B^{-1}$ and X. Their orders are $2, 2, 2, 2, 3$ and 4 respectively.

• (H2ii) A periodic element of $Aut(\bar{F}_2)$ is conjugate to exactly one of p, px, x^2, y^2, x and y. Their orders are $2, 2, 2, 3, 4$ and 6 respectively.

• (H3i) In $Aut(F_2)$, there are four conjugacy classes of elements of order 2, one of order 3, and one of order 4. The representatives of those of order 2 are α_1, α_2, α_3 and α_4; that of orders 3 and 4 are α_5 and α_6 respectively.

• (H3ii) In $Aut(\bar{F}_2)$, there are three conjugacy classes of elements of order 2, one of order 3, one of order 4, and one of order 6. The representatives of those of order 2 are $\bar{\alpha}_1$, $\bar{\alpha}_2$ and $\bar{\alpha}_3$; that of orders $3, 4$ and 6 are $\bar{\alpha}_5, \bar{\alpha}_6$ and $\bar{\alpha}_7$ respectively.

2.2.10 Remark.   • In the above Proposition, $\alpha_1 : x \mapsto y, y \mapsto x$, $\alpha_2 : x \mapsto x^{-1}, y \mapsto y^{-1}$, $\alpha_3 : x \mapsto x^{-1}, y \mapsto y$, $\alpha_4 : \alpha_3 \hat{x}$, $\alpha_5 : x \mapsto y, y \mapsto x^{-1} y^{-1}$, $\alpha_6 : x \mapsto y^{-1}, y \mapsto x$, $\alpha_7 : x \mapsto xy, y \mapsto x^{-1}$, and $\bar{\alpha}_i$s (for $1 \le i \le 7$) defined accordingly.

• It is pertinent to note that the above characterization does not necessarily hold for automorphisms of free groups of higher ranks $n > 2$.

2.3   Nielsen Automorphisms

2.3.1 Definition ([HMM05]). An automorphism $\theta \in Aut(F_X)$ is called a **Nielsen Automorphism** if for some $x \in X$,

$$\theta : x \mapsto x^{-1}, xy^{\pm 1} \text{ or } y^{\pm 1} x,$$

and for $y \in X$ (with $y \neq x$),

$$\theta : y \mapsto y.$$

2.3.2 Example. The Special automorphisms are key examples of Nielsen automorphisms.

2.3.3 Remark. Every Nielsen automorphism is a Whitehead automorphism, which we shall introduce in the next section.

2.3.4 Notation. We denote the set of all Nielsen automorphisms of F_n by ω_n.

2.3.5 Lemma ([HMM05]). $|\omega_n| = 4n(n-1)$.

2.4 The Whitehead Automorphisms

Here, we introduce a set of automorphisms which Whitehead ([Whi36a], [Whi36b]) used to deduce that if two elements of a finitely generated free group are equivalent under any automorphism of the group, then they are equivalent under a finite sequence of automorphisms which we shall refer to in the sequel as "Whitehead automorphisms". We start by defining some terms and notations that will be used.

2.4.1 Definition.   • For $w_1, w_2 \in F_n$, we denote **equivalence** by $\sim$, and write $w_1 \sim_\alpha w_2$ if $\alpha \in Aut(F_n)$ such that $\alpha(w_1) = w_2$.

• In F_X (where $\mid X \mid = n < \infty$), the set L_n of generators and inverses of F_n is defined as:

$$L_n = \{x_1, x_2, \cdots, x_n, \bar{x}_1, \bar{x}_2, \cdots, \bar{x}_n\} \text{ with } \bar{x}_i = x_i^{-1} \text{ for } 1 \le i \le n.$$

• A word $w \in F_n$ is said to be **Minimal** if $|w| \le |\alpha(w)| \ \forall \ \alpha \in Aut(F_n)$.

• A **Cyclic word** is a set of all cyclic permutations of a given cyclically reduced word.

• The **Length** of a cyclic word $w \in F_n$ is denoted by $|w|$, and defined as the number of letters in the cyclically ordered set.

2.4.2 Example (Using $F_2 = \langle x, y \rangle$).   • xx is minimal while yxx is not minimal as α_5 in Remark 2.2.10 reduces its length.

• An equivalence class of words in F_2 under $Aut(F_2)$ can be seen as:

$$\{xx, \overline{xx}, yy, \overline{yy}, xyxy, yxyx, x\bar{y}x\bar{y}, y\bar{x}y\bar{x}, \bar{x}y\bar{x}y, xyy\bar{x}, yxx\bar{y}, \bar{x}yyx, \bar{y}xxy, \cdots\}.$$

2.4.3 Problem (Whitehead). Suppose F is free on X; let $U = (u_1, u_2, \cdots, u_n)$ and $V = (v_1, v_2, \cdots, v_n)$ be finite ordered subsets of F_X.

• (I1) Is there an automorphism $\alpha : F_X \to F_X$ such that:

$$\alpha(u_i) = v_i \ \forall \ 1 \le i \le n?$$

• (I2) Is there an automorphism β which takes $\langle U \rangle$ into $\langle V \rangle$?

2.4.4 Remark. Lyndon and Schupp in [LS77] referred to the problems above as the Central Problems in the theory of automorphisms of free groups. They further gave a way of appreciating these problems in terms of matrices as seen in the Lemma below.

2.4.5 Lemma. Looking at U and V as matrices, the problem of "Is $\langle U \rangle = \langle V \rangle$?" can be viewed as "Do there exists a non-singular matrix P such that $V = PU$?". Problems (I1) and (I2) translates respectively to the following:

• (J1) Whether $V = UQ$ for some non-singular matrix Q?;

• (J2) Whether $V = PUQ$ for non-singular matrices P and Q?

2.4.6 Theorem (Whitehead). If $w_1, w_2 \in F_n$ such that $w_1 \sim w_2$, and w_2 is minimal, then there exists a sequence $\alpha_1, \alpha_2, \alpha_3, \cdots, \alpha_l$ of Whitehead automorphisms such that the following conditions are satisfied:

- (K1) $\alpha_l \alpha_{l-1} \alpha_{l-2} \cdots \alpha_2 \alpha_1 (w_1) = w_2$;

- (K2) $|\alpha_{i+1} \alpha_i \alpha_{i-1} \alpha_{i-2} \cdots \alpha_2 \alpha_1 (w_1)| \le |\alpha_i \alpha_{i-1} \alpha_{i-2} \cdots \alpha_2 \alpha_1 (w_1)|$ for $0 \le i \le (l-1)$, and with strict inequality unless $\alpha_i \alpha_{i-1} \alpha_{i-2} \cdots \alpha_2 \alpha_1 (w_1)$ is minimal.

We now define the Whitehead automorphisms referred in the Whitehead theorem above.

2.4.7 Definition (Whitehead Automorphisms).   • A **Type** 1 Whitehead Automorphism, α is a permutation which acts on elements of L_n, and preserves inverses as follows:

$$\alpha(\bar{x}) = \overline{\alpha(x)} \ \forall \ x \in L_n.$$

- A **Type** 2 Whitehead Automorphism, β is that which for a fixed $a \in L_n$ and $\beta(a) = a$, β carries each generator x of L_n (with $x \ne a, \bar{a}$) into one of $x, xa, \bar{a}x$ or $\bar{a}xa$ as follows:

$$\beta(x) = x, xa, \bar{a}x \text{ or } \bar{a}xa \ \forall \ x \in L_n \text{ (with } x \ne a, \bar{a}).$$

2.4.8 Notation.   • We denote the set of all Whitehead automorphisms of F_n by Ω_n.

- Let β be a Type 2 Whitehead automorphism as described in the definition above. We write $\beta = (A, a)$, where $A = \{a, y : \beta(y) = ya \text{ or } \beta(y) = \bar{a}ya\} \subseteq L_n$ with $y \ne a, \bar{a}$.
So, if $x \mapsto \bar{a}x$, then $\bar{x} \in A$, and if $x \mapsto \bar{a}xa$, then $x, \bar{x} \in A$.

2.4.9 Remark.   • As Type 1 automorphisms are permutations, they do not decrease the length of a word in F_n.

- In the Whitehead theorem, the automorphisms in condition (K2) can only be Type 2 Whitehead automorphisms (**the likely length decreasing ones**).

- We refer to either Type 1 or Type 2 Whitehead automorphisms as Whitehead automorphisms.

- It is now evident that the Nielsen automorphisms defined in Section 2.3 are examples of Whitehead automorphisms. So, $\psi_n \subset \omega_n \subset \Omega_n$.

Let us now appreciate Whitehead automorphisms with the following examples.

2.4.10 Example.   • In F_2, $\alpha : x_1 \mapsto \bar{x}_1, x_2 \mapsto \bar{x}_2$ is a Type 1 Whitehead automorphism while $\beta = (\{x_1, \bar{x}_2\}, x_1) : x_1 \mapsto x_1, \bar{x}_1 \mapsto \bar{x}_1, x_2 \mapsto \bar{x}_1 x_2, \bar{x}_2 \mapsto \bar{x}_2 x_1$ is a Type 2 Whitehead automorphism.

- In F_3, define $A = \{\bar{x}_1, \bar{x}_2, \bar{x}_3\}$. Then $\beta = (\{\bar{x}_1, \bar{x}_2, \bar{x}_3\}, \bar{x}_2) : x_1 \mapsto x_2 x_1, \bar{x}_1 \mapsto \bar{x}_1 \bar{x}_2, x_2 \mapsto x_2, \bar{x}_2 \mapsto \bar{x}_2, x_3 \mapsto x_2 x_3, \bar{x}_3 \mapsto \bar{x}_3 \bar{x}_2$ is a Type 2 Whitehead automorphism.

We give a simplified form of Type 2 Whitehead automorphism in the following Lemma:

2.4.11 Lemma. Let β be a Type 2 Whitehead automorphism (with $a \in L_n$ fixed) as described above. For each $x \in L_n$, the Whitehead Type 2 automorphism β acts on x as follows:

$$\beta(x) = (A, a)(x) := \begin{cases} x & \text{if } x, \bar{x} \notin A \\ xa & \text{if } x \in A \text{ and } \bar{x} \notin A \\ \bar{a}x & \text{if } x \notin A \text{ and } \bar{x} \in A \\ \bar{a}xa & \text{if } x, \bar{x} \in A. \end{cases}$$

2.4.12 Corollary. (A, a) never reduces length of a word w if both a and $\bar{a}$ are not in w.

As Type 2 Whitehead automorphisms are very important, one natural question is:

2.4.13 Question. Is the set of Type 2 Whitehead automorphisms finite?

The answer is contained in the following lemma.

2.4.14 Lemma ([MM04]). For any given F_n, there are $(2n4^{(n-1)} - 2n)$ non-trivial Type 2 Whitehead automorphisms.

2.4.15 Corollary. For $F_2 = \langle x_1, x_2 \rangle$, there are twelve non-trivial Type 2 Whitehead automorphisms: namely the ones given in table below.

A	(A, x_1)	$(A, \bar{x}_1)$	(A, x_2)	$(A, \bar{x}_2)$
$\{x_1, x_2\}$	$x_1 \mapsto x_1,\ \bar{x}_1 \mapsto \bar{x}_1,\ x_2 \mapsto x_2 x_1,\ \bar{x}_2 \mapsto \bar{x}_1 \bar{x}_2$		$x_1 \mapsto x_1 x_2,\ \bar{x}_1 \mapsto \bar{x}_2 \bar{x}_1,\ x_2 \mapsto x_2,\ \bar{x}_2 \mapsto \bar{x}_2$	
$\{\bar{x}_1, x_2\}$		$x_1 \mapsto x_1,\ \bar{x}_1 \mapsto \bar{x}_1,\ x_2 \mapsto x_2 \bar{x}_1,\ \bar{x}_2 \mapsto x_1 \bar{x}_2$	$x_1 \mapsto \bar{x}_2 x_1,\ \bar{x}_1 \mapsto \bar{x}_1 x_2,\ x_2 \mapsto x_2,\ \bar{x}_2 \mapsto \bar{x}_2$	
$\{x_1, \bar{x}_2\}$	$x_1 \mapsto x_1,\ \bar{x}_1 \mapsto \bar{x}_1,\ x_2 \mapsto \bar{x}_1 x_2,\ \bar{x}_2 \mapsto \bar{x}_2 x_1$			$x_1 \mapsto x_1 \bar{x}_2,\ \bar{x}_1 \mapsto x_2 \bar{x}_1,\ x_2 \mapsto x_2,\ \bar{x}_2 \mapsto \bar{x}_2$
$\{\bar{x}_1, \bar{x}_2\}$		$x_1 \mapsto x_1,\ \bar{x}_1 \mapsto \bar{x}_1,\ x_2 \mapsto x_1 x_2,\ \bar{x}_2 \mapsto \bar{x}_2 \bar{x}_1$		$x_1 \mapsto x_2 x_1,\ \bar{x}_1 \mapsto \bar{x}_1 \bar{x}_2,\ x_2 \mapsto x_2,\ \bar{x}_2 \mapsto \bar{x}_2$
$\{x_1, \bar{x}_1, x_2\}$			$x_1 \mapsto \bar{x}_2 x_1 x_2,\ \bar{x}_1 \mapsto \bar{x}_2 \bar{x}_1 x_2,\ x_2 \mapsto x_2,\ \bar{x}_2 \mapsto \bar{x}_2$	
$\{x_1, \bar{x}_1, \bar{x}_2\}$				$x_1 \mapsto x_2 x_1 \bar{x}_2,\ \bar{x}_1 \mapsto x_2 \bar{x}_1 \bar{x}_2,\ x_2 \mapsto x_2,\ \bar{x}_2 \mapsto \bar{x}_2$
$\{x_1, x_2, \bar{x}_2\}$	$x_1 \mapsto x_1,\ \bar{x}_1 \mapsto \bar{x}_1,\ x_2 \mapsto \bar{x}_1 x_2 x_1,\ \bar{x}_2 \mapsto \bar{x}_1 \bar{x}_2 x_1$			
$\{\bar{x}_1, x_2, \bar{x}_2\}$		$x_1 \mapsto x_1,\ \bar{x}_1 \mapsto \bar{x}_1,\ x_2 \mapsto x_1 x_2 \bar{x}_1,\ \bar{x}_2 \mapsto x_1 \bar{x}_2 \bar{x}_1$		

2.4.16 Notation. As noted earlier, Ω_n denotes the set of all Whitehead automorphisms of F_n. We now denote the set of Whitehead automorphisms of Types 1 and 2 by $_1\Omega_n$ and $_2\Omega_n$ respectively. So, $\Omega_n = {_1\Omega_n} \cup {_2\Omega_n}$. We sometimes write Ω_n as Ω to avoid ambiguity with index notations.

2.4.17 Lemma. $|_1\Omega_n| = 2^n n!$.

2.4.18 Corollary. $|\Omega_n| < \infty$.

2.4.19 Notation. Given a word $w \in F_n$.

- We denote the set of Type 2 Whitehead automorphisms that do not reduce the length of w as described in Corollary 2.4.12 by $_2\underline{\Omega_n}(w)$ (or $_2\underline{\Omega_n}$ for short), and call it the **Bad Type** 2 **Whitehead Automorphisms**.

- Similarly, the complement of $_2\underline{\Omega_n}$ in $_2\Omega_n$ will be denoted by $_2\overline{\Omega_n}$, and called the **Maybe Good Type** 2 **Whitehead Automorphisms**.

- A subset of $_2\overline{\Omega_n}$ consisting of only the length reducing Type 2 Whitehead automorphisms will be called the **Good Type** 2 **Whitehead Automorphisms**, and denoted by $_2\overline{\overline{\Omega_n}}$.

2.4.20 Remark.   • The Good Type 2 Whitehead automorphisms are simply the length reducing ones while the Bad Type 2 Whitehead automorphisms are the particular non-length reducing ones identified in Corollary 2.4.12. On the other hand, $_2\overline{\Omega_n}$ may contain both reducing and non-reducing length Type 2 Whitehead automorphisms for any given word.

- Each non-minimal word of length 2 in F_2 has exactly two Good Type 2 Whitehead automorphisms. The second Good Whitehead automorphism can always be gotten whenever the first one is known. For example, if the first Good Type 2 Whitehead automorphism fixes the first generator and maps the second to the product of the first and second, then the second Good Type 2 Whitehead automorphism will fix the second generator, and map the first to product of second and first...with the inverse operation interchanged.

- In theory, it is not easy to find all the Good Whitehead automorphisms for a particular non-minimal word in free groups of finite ranks $n \geq 4$. However, this is a very easy task in programming.

2.5   Equivalence of Elements under $Aut(F_n)$

In this section, we investigate the equivalence of elements under $Aut(F_n)$ following [HL74] and [LS77]. We start by introducing the main results of this study.

2.5.1 Proposition ([HL74]). Let u, v and w be cyclic words such that $\alpha(w) = u$ and $\gamma(w) = v$ for $\alpha, \gamma \in \Omega$, with $|u| \leq |w|$ and $|v| \leq |w|$, and either $|u| < |w|$ or $|v| < |w|$. Then there exist $\alpha_1, \alpha_2, \cdots, \alpha_n \in \Omega$ such that:

$$\alpha_n \alpha_{n-1} \alpha_{n-2} \cdots \alpha_2 \alpha_1(u) = v \ (n \geq 0),$$

and defining $\alpha_l \alpha_{l-1} \alpha_{l-2} \cdots \alpha_2 \alpha_1(u) = u_l$ (with $0 \leq l \leq n$), we have:

$$|u_l| < |w| \ \forall \ l \in (0, n).$$

2.5.2 Theorem (Restricted Form of Whitehead's Theorem). If w and w' are cyclic words such that $w \sim_{\alpha \in Aut(F_n)} w'$ (for $w, w' \in F_n$), and w' is minimal for their equivalence class, then there exist $\alpha_1, \alpha_2, \cdots, \alpha_n \in \Omega$ (with $n \geq 0$) such that $\alpha_l \alpha_{l-1} \alpha_{l-2} \cdots \alpha_2 \alpha_1(w) = w_l$ (with $0 \leq l \leq n$) imples that:

$$w_n = w',$$

and

$$|w_m| \leq |w| \ \forall \ m \in [1, n],$$

with strict inequality unless w is minimal.

As only Type 2 Whitehead automorphisms can decrease the length of a word, to appreciate the above results, we explore other tools involving Type 2 Whitehead automorphisms.

2.5.3 Lemma. Let (A, a) be a Type 2 Whitehead automorphism. Then the following holds.

- (L1) $(A, a)^{-1} = (A - a + \bar{a}, \bar{a})$.

- (L2) If A' is the complement of A in L_n, then $(A', \bar{a})^{-1} = (A' - \bar{a} + a, a)$.

- (L3) $(A, a)(A', \bar{a})^{-1} = (A, a)(A' - \bar{a} + a, a)$ is the inner automorphism $(L_n - \bar{a}, a)$ which takes each element $w \in F_n$ into $\bar{a}wa$.

- (L4) Let $w \in F_n$ be a cyclic word. As inner automorphisms have no effect on Cyclic words, $(A, a)(w) = (A', \bar{a})(w)$.

2.5.4 Definition. Given $A, B \subseteq L_n$, and a cyclic word $w \in F_n$. We define $(A.B)_w$ (or $A.B$ for short) as the number of consecutive pairs of letters of the form $x\bar{y}$ or $y\bar{x}$ in w (for $x \in A$ and $y \in B$).

2.5.5 Notation. For $A \cap B = \varnothing$, we use "$A + B$" for $A \cup B$, and "$A - B$" for $A \cap B'$.

For ease of reference, we shall refer to the $A.B$ operation defined above as **Character pair**.

Below are some basic operations associated with Character pairs.

2.5.6 Lemma.   • (M1) $A.B \geq 0$;

- (M2) $A.B = B.A$;

- (M3) $(A + B).C = A.C + B.C$ and $(A - B).C = A.C - B.C$;

- (M4) $a.a = 0$;

- (M5) $a.L_n = \bar{a}.L_n$ is the total number of letters a and $\bar{a}$ in w.

2.5.7 Definition. Let $\beta = (A, a)$ be a Type 2 Whitehead automorphism. For a cyclic word w, we define $\Delta_w(\beta)$ (or $\Delta(\beta)$ for short) as follows:

$$\Delta(\beta) = |\beta(w)| - |w|.$$

2.5.8 Proposition. $\Delta(\beta) = A.A' - a.L_n$.

To prove this proposition, let us first prove the following lemma.

2.5.9 Lemma. Let w_1 be a non-reduced word gotten from w by replacing each letter x with $\beta(x)$, and not deleting any subword of the form $a\bar{a}$ or $\bar{a}a$. Suppose w_2 is the resulting term after deleting all subwords of the form $a\bar{a}$ or $\bar{a}a$ from w_1. Then w_2 is reduced.

[For ease of reference, we call the process of passing w to w_1 the **transformation process**.]

Proof.   • As w is reduced, and w_1 is gotten from w by introducing either a or $\bar{a}$, we expect that w_1 may contain subwords of the form $a\bar{a}$ or $\bar{a}a$.

- But a new letter a can only be introduced in the transformation process only if it is following an old x (with $x \in A - a$, or simply say $x \neq a, \bar{a}$). So, such a never follows $\bar{a}$. In the same vein, a new letter $\bar{a}$ can only be introduced in the transformation process only if it is following an old y (with $y \neq a, \bar{a}$). Hence, w_1 cannot contain any subword of the form $\bar{a}a$.

- Now, to prove that w_2 is reduced suffices to show that w_2 cannot contain any subword of the form $ya\bar{a}x$.

- Suppose w_1 contains a non-reduced subword of the form $ya\bar{a}x$. Then, by deleting $a\bar{a}$, we have a reduced word yx in w_2. Suppose only one of the letters in a subword $a\bar{a}$ (say a) was intially present in w, then yax transforms to either $ya\bar{a}x$ or $yaa\bar{a}x$ in w_1; and further into yx or yax in w_2, according as $y \neq \bar{x}$ or $y \neq \bar{a}$. Thus, w_2 is reduced.

$\square$

A consequence of Lemma 2.5.9 is:

2.5.10 Corollary. $\beta(w) = w_2$.

2.5.11 Lemma. Let $\beta = (A, a)$ be a Type 2 Whitehead automorphism as decribed above. Then

$$\Delta(\beta) = \Delta_1 - \Delta_2, \tag{2.5.1}$$

where Δ_1 is the number of letters a or $\bar{a}$ introduced in the transformation process and remains part of w_2, and Δ_2 is the number of letters a or $\bar{a}$ introduced in the transformation process and do not remain part of w_2.

We now prove Proposition 2.5.8.

Proof. • A newly introduced letter "a" following $x \in A - a$ will not be followed in w_1 by an $\bar{a}$ if and only if x occurs in a subword $x\bar{y}$ of w for $y \in A'$. Similarly, a newly introduced letter "$\bar{a}$" following $\bar{x} \in A - a$ will not be followed in w_1 by an a if and only if $\bar{x}$ occurs in a subword $y\bar{x}$ of w for $y \in A'$. Thus,

$$\Delta_1 = (A - a).A'. \tag{2.5.2}$$

- Also, a newly introduced letter "a" following $x \in A - a$ will be followed in w_1 by an initially present letter $\bar{a}$ if and only if x occurs in a subword $x\bar{a}$ of w. Similarly, a newly introduced letter "$\bar{a}$" following $\bar{x} \in A - a$ will be followed in w_1 by an initially present letter a if and only if $\bar{x}$ occurs in a subword $a\bar{x}$ of w. Thus,

$$\Delta_2 = (A - a).a. \tag{2.5.3}$$

- Now,

$$
\begin{aligned}
\Delta(\beta) &= \Delta_1 - \Delta_2 \ \text{(From 2.5.1)} \\
&= (A - a).A' - (A - a).a \ \text{(From 2.5.2 and 2.5.3)} \\
&= A.A' - a.A' - A.a + a.a \ \text{(By (M3))} \\
&= A.A' - a.A' - a.A \ \text{(By (M4) and (M2))} \\
&= A.A' - a.(A' + A) \ \text{(By (M3))} \\
&= A.A' - a.L_n \ \text{(Since } A \cap A' = \varnothing).
\end{aligned}
$$

$\square$

We now visit an important result which Higgins and Lyndon in [HL74] used to finalise the proof of Proposition 2.5.1. It is essential because it gives an upper bound on the finite set of Whitehead automorphisms. Many thanks to Higgins and Lyndon !!!

2.5.12 Lemma. Suppose w is a fixed cyclic word. Let $\alpha(w) = u$ and $\gamma(w) = v$ for $\alpha, \gamma \in \Omega$, with $|u| \leq |w|$ and $|v| \leq |w|$, and either $|u| < |w|$ or $|v| < |w|$. Then:

- (N1) $|w| > \frac{1}{2}(|u| + |v|)$;

- (N2) $\exists \, \alpha_1, \alpha_1, \cdots \alpha_n \in \Omega$ (with $n \leq 4$) such that:

$$\alpha_n \alpha_{n-1} \alpha_{n-2} \cdots \alpha_2 \alpha_1(u) = v$$

and
$$|\alpha_l \alpha_{l-1} \alpha_{l-2} \cdots \alpha_2 \alpha_1(u)| < |w| \; \forall \, l \in (0, n).$$

[LS77] extended Theorem 2.5.2 to a finite set of cyclic words as shown in the following Proposition.

2.5.13 Proposition. Let $w_1, w_2, \cdots, w_r$ and $w_1', w_2', \cdots, w_r'$ be two sequences of cyclic words such that $\alpha(w_1) = w_1', \alpha(w_2) = w_2', \cdots, \alpha(w_r) = w_r'$ for some $\alpha \in Aut(F_n)$. If $\sum |w_h'|$ is minimal among all $\sum |\alpha'(w_h)|$ for $\alpha' \in Aut(F_n)$, then there exist $\alpha_1, \alpha_2, \cdots, \alpha_n \in \Omega$ (with $n \geq 0$) such that $\alpha_n \alpha_{n-1} \cdots \alpha_2 \alpha_1 = \alpha$ implies that:

$$\sum |\alpha_l \alpha_{l-1} \cdots \alpha_2 \alpha_1(w_h)| \leq \sum |w_h| \text{ for } l \in (0, n),$$

with strict inequality unless $\sum |w_h|$ is minimal.

We come to the end of this section by appreciating what we have studied using the following Lemma.

2.5.14 Lemma. If $\underline{w_1}$ and $\underline{w_2}$ are two elements of a finitely generated free group, then it is decidable whether there is an automorphism $\alpha \in Aut(F_n)$ such that $\alpha(\underline{w_1}) = \underline{w_2}$.

Proof.   • Let w_1 and w_2 be the cyclic words corresponding to $\underline{w_1}$ and $\underline{w_2}$ respectively.

- For each $\alpha \in \Omega$, check whether $\alpha(w_1) = w_1'$ with $|w_1'| < |w_1|$ for $w_1' \in F_n$. Also check whether $\alpha(w_2) = w_2'$ with $|w_2'| < |w_2|$ for $w_2' \in F_n$.

- Substitute w_j' for w_j ($j = 1, 2$) if any of the above cases is true, and repeat that step until we get w_j^{min}, where w_j^{min} is the minimal word equivalent to w_j for such j.

- By Corollary 2.4.18, the above process must terminate.

- As both w_1^{min} and w_2^{min} are minimal, such automorphism can only exist if $|w_1^{min}| = |w_2^{min}|$.

- Suppose $|w_1^{min}| = |w_2^{min}|$.

- By Lemma 1.2.6, the set of cyclic words of a given length l is finite.

- Let us take this set of cyclic words as the set of vertices $V(G)$ of a graph G, such that for $w', w'' \in V(G)$, an edge is initiated from w' to w'' whenever there exists a Whitehead automorphism from w' to w''.

- By Theorem 2.5.2, there exists an automorphism $\alpha : w_1^{\min} \mapsto w_2^{\min}$ whenever there is a connected path from $w_1^{\min}$ to $w_2^{\min}$. In such case, we say $w_1^{\min} \sim w_2^{\min}$; and hence, $\underline{w_1} \sim \underline{w_2}$.

$\square$

Lemma 2.5.14 can be extended to a finite set of cyclic words as follows:

2.5.15 Lemma. If $w_1, w_2, \cdots, w_r$ and $w_1', w_2', \cdots, w_r'$ are two sequences of cyclic words in a finitely generated free group, then it is decidable whether there is an automorphism $\alpha \in Aut(F_n)$ such that $\alpha(w_1) = w_1', \alpha(w_2) = w_2', \cdots, \alpha(w_r) = w_r'$.

Proof. Similar to the proof of Lemma 2.5.14 above. $\square$

2.6   A Presentation for $Aut(F_n)$

We begin with a review of works done by Rapaport (in [Rap58]) before following McCool in [McC74] to give a presentation for $Aut(F_n)$. Our choice of McCool's approach is owning to the fact that it helps us to appreciate the beauty of Whitehead automorphisms introduced earlier.

Rapaport gave a graphical way of appreciating length of a word w and $\alpha(w)$ for an automorphism α. She then used it to explain the following theorem:

2.6.1 Theorem ([Rap58]). Let $A = \alpha_2 \alpha_1$ such that $|\alpha_1(w_0)| > |w_0|$ and $|Aw_0| \leq |\alpha_1(w_0)|$, where w_0 is a cyclic word in F_n. Then, there exists a factorization $B = B_l B_{l-1} \cdots B_1$ of A such that for every intermediate word $w_k' = B_k B_{k-1} \cdots B_1(w_0)$ (with $k < l$), $|w_k'| < |\alpha_1(w_0)|$, where B_i's are T-transformations or level transformations.

In giving a corollary to the above theorem, Rapaport pointed out that the theorem also holds if w_0 is a set of finite number of words. We now give a brief description of notations used in the theorem, and conclude Rapaport's discussion by presenting some consequences of the theorem.

2.6.2 Notation. Let a, b, c, z denote fixed subsets of the generators of F_n; and let d denote a generator or the inverse of a generator such that the sets $a, b, c, d^{\pm 1}$ are disjoint, and the set (a, b, c, z) contains every generator a_i of F_n exactly once.

2.6.3 Definition.   • For every subdivision of the generators of F_n as in Notation 2.6.2, a T- **transformation** is a transformation of the form:

$$T : a \mapsto ad, b \mapsto \bar{d}b, c \mapsto \bar{d}cd, z \mapsto z.$$

- For two T- transformations T_1 and T_2 such that $T_1(a_i) = v_i(a_1, \cdots a_n) = v_i(a)$ and $T_2(a_i) = w_i(a_1, \cdots, a_n) = w_i(a)$, we define the **product** $T_2 T_1$ (where $T_2 T_1(a_i) = v_i(w_1(a), \cdots, w_n(a))$ as follows:

$$T_2 T_1 : a \mapsto ad, b \mapsto \bar{d}b, c \mapsto \bar{d}cd, a' \mapsto a'd', b' \mapsto \bar{d}'b', c' \mapsto \bar{d}'c'd',$$

with the subdivisions $(a, \cdots)$ and $(a', \cdots)$ of generators of F_n, and $z \mapsto z, z' \mapsto z'$ omitted.

- A T- transformation α is called a **level transformation on a word** $w \in F_n$ if $|\alpha(w)| = |w|$.

- A T- transformation α is called a **level transformation** if it is a level transformation $\forall \, w \in F_n$.

2.6.4 Example. Type 1 Whitehead automorphisms are level transformations. On the other hand, the Good Type 2 Whitehead automorphisms are not level transformations.

Below are some consequences of Theorem 2.6.1.

2.6.5 Lemma.   • (O1) If w_0 is a set of elements, and α is an automorphism of F_n with $\alpha(w_0) = w$, then there exist T- transformations B_i $(1 \leq i \leq k)$ with

$$\Pi_{i=1}^k B_i = \alpha$$

such that every set of words $\Pi_{i=1}^r B_i(w_0)$ (for $r \leq k$) is at most as long as $\max(|w_0|, |w|)$.

- (O2) Words that are minimal with respect to all T- transformations are also minimal with respect to any other automorphism of F_n.

- (O3) Suppose there exist an automorphism α such that $w_2 = \alpha(w_1)$ is minimal, where $w_1, w_2 \in F_n$. Then, the number of distinct generators in w_1 can be decreased by applying a transformation if and only if w_2 has fewer generators than w_1.

Proof. See pages 160 and 161 of [Rap58].   □

We now follow McCool in [McC74] to give a presentation for $Aut(F_n)$.

First and foremost, we recall the following basic definition.

2.6.6 Definition. A presentation $\langle X | R \rangle$ of a group G is said to be **finite** if the set of generators X and the set of relators R are finite. We say that such group is finitely presented.

2.6.7 Theorem (Nielsen 1924). $Aut(F_n)$ is finitely presented.

Before 1974, Theorem 2.6.7 was known. However, McCool gave an alternative presentation P_n for $Aut(F_n)$ as contained in the following proposition.

2.6.8 Proposition. The set X of generators of P_n is the set of all Whitehead automorphisms, and the relations R are all possible relations of the forms given below:

- (P1) $(A, a)^{-1} = (A - a + \bar{a}, \bar{a})$;

- (P2) $(A, a)(B, a) = (A \cup B, a)$ if $A \cap B = \{a\}$;

- (P3) $(B, b)^{-1}(A, a)(B, b) = \begin{cases} (A, a) & \text{if } A \cap B = \varnothing, \bar{a} \notin B, \text{ and } \bar{b} \notin A \\ (A + B - b, a) & \text{if } A \cap B = \varnothing, \bar{a} \notin B, \text{ and } \bar{b} \in A; \end{cases}$

- (P4) $(A, a)(A - a + \bar{a}, b) = \left(\begin{smallmatrix} a & b \\ b & a \end{smallmatrix} \right)(A - b + \bar{b}, a)$ if $b \in A, \bar{b} \notin A$ and $a \neq b$ (where $\left(\begin{smallmatrix} a & b \\ b & a \end{smallmatrix} \right)$ is the automorphism : $a \mapsto \bar{b}, b \mapsto a$, and fixes all other letters $x \notin \{a, \bar{a}, b, \bar{b}\}$);

- (P5) $T^{-1}(A, a)T = (AT, aT)$ if T is a Type 1 Whitehead automorphism;

- (P6) Consists of any set of defining relations for $_1\Omega_n$ on the generating set consisting of Type 1 Whitehead automorphisms.

2.6.9 Notation. We denote a presentation obtained from P_n by removing (P6) relations by P'_n.

2.6.10 Lemma. The relations in P'_n are:

- (Q1) $(A, a) = (L_n - \bar{a}, a)(A', \bar{a}) = (A', \bar{a})(L_n - \bar{a}, a)$;

- (Q2) $(L_n - b, \bar{b})(A, a)((L_n - b, \bar{b}) = (A, a)$ if $b, \bar{b} \in A'$;

- (Q3) $(L_n - b, \bar{b})(A, a)(L_n - \bar{b}, b) = (A', \bar{a})$ if $b \neq a, b \in A$ and $\bar{b} \in A'$.

2.6.11 Definition. Let $\alpha \in \Omega_n$. An element $x_n \in L_n$ is **not used in** α if either

- α is a Type 1 Whitehead automorphism such that $\alpha(x_n) = x_n$ **or**

- $\alpha = (A, a)$ is a Type 2 Whitehead automorphism such that $x_n, \bar{x}_n \notin A$.

2.6.12 Remark. If x_n is not used in α, then α is seen as an element of Ω_{n-1}.

We conclude this section by visiting two results McCool established while proving Proposition 2.6.8, starting with a result he used to appreciate Proposition 2.5.1 as well as obtain a refinement of the Corollary to Theorem 2.6.1.

2.6.13 Proposition. Let $\alpha(w) = u$ and $\gamma(w) = v$, where u, v and w are cyclic words; $\alpha, \gamma \in \Omega_n$, with $|u| \leq |w|$ and $|v| \leq |w|$, and either $|u| < |w|$ or $|v| < |w|$. Then there exist $\alpha_1, \alpha_2, \cdots, \alpha_r \in \Omega_n$ (with $r \leq 4$) such that

- (R1) $\alpha_r \alpha_{r-1} \cdots \alpha_1(u) = v$,

- (R2) $|\alpha_l \alpha_{l-1} \cdots \alpha_1(u)| < |w|$ for $l \in (0, r)$,

- (R3) $\alpha^{-1} \gamma = \alpha_r \alpha_{r-1} \cdots \alpha_1$,

- (R4) the relation (R3) holds in P'_n,

- (R5) if x_n is used in α and γ, then x_n is not used in $\alpha_1, \alpha_2, \cdots, \alpha_r$. The relation (R3) holds in P_{n-1} if $\alpha, \gamma, \alpha_1, \cdots, \alpha_{r-1}, \alpha_r$ are elements of Ω_{n-1}.

Proof. See pages 262–266 of [McC74]. □

Another beautiful result (due to McCool) is given as follows:

2.6.14 Lemma. Let $U = (u_1, u_2, \cdots, u_m)$, $V = (v_1, v_2, \cdots, v_m)$ and $W = (w_1, w_2, \cdots, w_m)$ be sequences of elements of F_n; and let α and γ be elements of Ω_n such that $\alpha(W) = U$ and $\gamma(W) = V$ with $|U| \leq |W|$, $|V| \leq |W|$, and either $|U| < |W|$ or $|V| < |W|$. Then, there exist $\alpha_1, \cdots, \alpha_r \in \Omega_n$ (with $r \leq 4$) such that the following holds:

- (S1) $\alpha^{-1} \gamma = \alpha_r \cdots \alpha_1$;

- (S2) the relation (S1) holds in P'_n;

- (S3) $|\alpha_l \cdots \alpha_1(U)| < |W|$ for $0 < l < r$.

Proof. See pages 261 and 262 of [McC74]. □

2.6.15 Remark. Though McCool's method of obtaining a presentation for a finitely generated free group uses Whitehead automorphisms as generators, it is pertinent to note that these generators have infinite orders. Nielsen [Nie24] and Gersten [Ger84] also used generators of infinite orders. To see a presentation whose generators have finite orders, see [Neu32]. Armstrong, Forrest and Vogtmann also gave a beautiful presentation for $Aut(F_n)$. See [AFV08] for further details.

3. The Whitehead Algorithm

We seek to design algorithms for appreciating results from the previous chapter as well as Whitehead's result for finitely generated free groups using [GAP13]. We start with a re-examination of the Whitehead's theorem.

3.1 Whitehead Theorem Revisited

Thanks to Whitehead for giving us a theorem that enables us know of existence of a finite sequence of automorphisms such that if two elements of a finitely generated free group are equivalent under any automorphism of the group, then they are equivalent under such finite sequence of automorphisms. Whitehead also deduced that the lengths of the words gotten after applying each Whitehead automorphism are monotonistically strictly decreasing until minimality is attained.

3.1.1 Need for Whitehead Algorithm. From the Whitehead's theorem, and its consequences studied in the last chapter, it is evident that the Whitehead algorithm is a powerful tool for determining equivalence classes of minimal words in a finitely generated free group.

We now introduce terms that we will use from time to time.

3.1.2 Definition. • A word $w \in F_n$ is said to be **Whitehead reducible** if it is non-minimal; i.e. if there exist a type 2 Whitehead automorphism β such that $|\beta(w)| < |w|$. On the other hand, w is called **Whitehead irreducible** if it is not Whitehead reducible.

• A word $w \in F_n$ is said to be **Whitehead equivalent** to another word $w_1 \in F_n$ if there exist Whitehead automorphisms $\gamma_1, \cdots, \gamma_{k-1}, \gamma_k$ such that $\gamma_k \gamma_{k-1} \cdots \gamma_1(w) = w_1$.

From now onwards, whenever we mention reducible(irreducible), we mean Whitehead reducible(irreducible). Similar statement holds for equivalence.

3.1.3 Lemma. Two irreducible words of different lengths are not equivalent.

Proof. • Let w and w_1 be two irreducible words of different lengths. W.L.O.G, suppose w_1 is gotten from w by Whitehead equivalence, and $|w| > |w_1|$.

• As w and w_1 are of different lengths, by Remark 2.4.9 and Notation 2.4.19, the automorphism that induces the equivalence must involve a good type 2 Whitehead automorphism.

• This further implies that w is reducible; thus contradicting the hypothesis that w is irreducible.

$\square$

3.2   Algorithm Developments

A lot has been said about Whitehead automorphisms in the previous chapter. So, we skip the algorithm for it. However, we shall design programs aimed at appreciating it in the next section. Whitehead's theorem induces many algorithms for which a couple of them will be discussed below.

3.2.1 Algorithm. For constructing a function that will be used to check whether a word $w \in F_n$ is minimal or not.

- Step 0. Set the input of the proposed function "IsMin" as w and F_n.

- Step 1. Start with found=true; then set found=false if $|\beta(w)| < |w|$ for any $\beta \in {}_2\Omega_n$ [To reduce computation time, you may wish to take β from $\overline{{}_2\Omega_n}$.]

- Step 2. Return true or false according as found=true or found=false.

3.2.2 Algorithm. For finding all minimal words of length l in F_n.

- Step 1. Find all the reduced words of length l in $G = F_n$, and call it R_{nl}. By Lemma 1.2.6, this is possible.

- Step 2. Create an empty list of minimal words of length l in G (Say $M_{nl} := [\ ];$).

- Step 3. For each $w \in R_{nl}$, use IsMin function constructed in Algorithm 3.2.1 to check whether w is minimal or not, and append w to the list M_{nl} whenever the evaluation is true. The list gotten from this step is the desired list.

3.2.3 Algorithm. For finding a minimal word equivalent to a word $w \in F_n$.

- Step 1. Use the IsMin function constructed in Algorithm 3.2.1 to check whether or not w is minimal. If w is minimal, return w; else proceed to Step 2.

- Step 2. Substitute $\beta(w)$ for w whenever $|\beta(w)| < |w|$ for $\beta \in {}_2\Omega_n$.

- Step 3. Repeat Step 2 until no further reduction is obtainable. Return the irreducible word obtained from this step.

3.2.4 Algorithm. For finding a list of minimal words equivalent to a word $w \in F_n$, and collecting all the Whitehead automorphisms used in this process.

- Step 1. Create an empty list of Whitehead automorphisms (Say WA:=[ ];).

- Step 2a. Apply Steps 1 to 3 of Algorithm 3.2.3 (where necessary), and denote the resulting word by w^{last}.

- Step 2b. For non-minimal case, append any β gotten from Steps 2 and 3 of Algorithm 3.2.3 to the list WA.

- Step 3. Create a singleton list MinEquivElts containing w^{last}.

- Step 4a. For each $\beta \in {}_2\Omega_n$, if $\beta(w^{\text{last}}) = \underline{w}$, $|\underline{w}| = |w^{\text{last}}|$ and $\underline{w}$ is not already in the list MinEquivElts, then append $\underline{w}$ to MinEquivElts.

- Step $4b$. Append any β used in Step $4a$ to the list WA.

- Step $5a$. For each $\alpha \in {}_1\Omega_n$, if $\alpha(w^{\text{last}}) = \underline{w}$, $|\underline{w}| = |w^{\text{last}}|$ and $\underline{w}$ is not already in the list MinEquivElts, then append $\underline{w}$ to MinEquivElts.

- Step $5b$. Append any α used in Step $5a$ to the list WA.

- Step $6a$. Print("The list of all minimal words equivalent to w is", MinEquivElts) where MinEquiv-Elts is the final list obtained after applying Step $5a$.

- Step $6b$. Print("The Whitehead automorphisms used in this collection are", WA).

3.2.5 Algorithm. For determining whether two elements of a finitely generated free group are equivalent.

- Step 0. Take $w, w_1 \in F_n$.

- Step $1a$. Use Algorithm 3.2.3 to find a minimal word equivalent to w.

- Step $1b$. Use Algorithm 3.2.4 to find all minimal words equivalent to w_1.

- Step 2. Return true if the answer in step $1a$ is contained in the answer in step $1b$; and false if otherwise.

3.2.6 Remark. Algorithm 3.2.5 can be used to classify all minimal words of a certain length l in F_n up to equivalence. We shall investigate this in the next section.

3.3   GAP Exhibitions

Here, we appreciate the beauty of this study by designing series of codes using Groups, Algorithms and Programming (GAP). We begin by designing programs that will be used to completely understand Whitehead's automorphisms, then test for minimality, list all minimal words equivalent to a word $w \in F_n$, investigate whether two words in a finitely generated free group are equivalent or not, and lastly classify all minimal words of a certain length l in F_n up to equivalence. We carry out this task using [GAP13].

3.3.1 Program. A program for finding all Type 1 Whitehead automorphisms of F_n.

```
WA1:= function (F)
# This function takes a finitely generated free group as F.
# It returns all Type 1 Whitehead automorphisms of F.
local gens, n, S, h, l, auts, ims, C, j, c, k, y;
gens:= GeneratorsOfGroup(F); n:=Length(gens);
S:=SymmetricGroup(n);
h:=Orbit(S,gens,Permuted); l:=Length(h);
auts:=[]; ims:=[];
C:=Cartesian(List([1..n], x->[-1,1]));
for j in [1..l] do
    for c in C do
        for k in [1..n] do
            y:=c[k];
            ims[k]:=h[j][k]^y;
```

```
        od ;
        # We now append the automorphisms to the list auts.
        Add( auts , GroupHomomorphismByImages( F , F , gens , ims ) );
    od ;
od ;
return auts ;
end ;
```

3.3.2 Program. A program for finding all Type 2 Whitehead automorphisms of F_n.

```
WA2:= function ( F )
# This function takes a finitely generated free group as F.
# It returns all Type 2 Whitehead automorphisms of F.
local gens , n , C , auts , i , j , e , a , ims , c , cct ;
gens:= GeneratorsOfGroup ( F ); n:= Length ( gens );
# Forming Cartesian product of (n−1) copies of [1,2,3,4] to
#help loop through automorphisms for a fixed generator a.
C:= Cartesian ( List ( [1 .. n−1], x − >[1,2,3,4])); auts := [];
for i in [1 .. n] do for e in [−1,1] do
    a := gens[ i ]^e;
    for c in C do if c <> List ([1 .. n−1], x−>1) then
        ims := []; cct := 0;
        for j in [1 .. n] do
            if j = i then
                ims[ j ] := gens[ i ];
            else
                cct := cct + 1;
                # The image of gens[j] will be one of gens[j], a^−1*gens[j],
                #F.j*gens[j] or a^−1*gens[j]*a when c[cct]=1,2,3 or 4 resp.
                if c[ cct ] = 1 then ims[ j ] := gens[ j ];
                elif c[ cct ] = 2 then ims[ j ] := gens[ j ] * a;
                elif c[ cct ] = 3 then ims[ j ] := a^−1 * gens[ j ];
                else ims[ j ] := a^−1 * gens[ j ] * a;
                fi ;
            fi ;
        od ;
        # We now append the automorphism this defines to the list auts.
        Add( auts , GroupHomomorphismByImages( F , F , gens , ims ) );
    fi ; od ;
od ; od ;
return auts ;
end ;
```

3.3.3 Program. A program for checking whether a word in F_n is minimal or not.

```
IsMin:=function(w,G)
#  IsMin  takes  a  finitely  generated  free  group  G,  and  a  word  w  in  G.
#  It  returns  true  if  w  is  minimal,  and  false  if  otherwise.
local  wa2,  b,  found;
wa2:=WA2(G);  #  This  is  the  list  of  Type  2  Whitehead  Automorphisms.
for  b  in  wa2  do
     found:=true;
     if  Length(Image(b,w))<Length(w)  then
          found:=false;
          break;
     fi;
od;
if  found=false  then
     return  false;
else
     return  true;
fi;
end;
```

See Appendix A.1 for a program for finding all reduced words of lengths l in F_n. Appendix A.2 contains a program which can be used to find all minimal words of lengths l in F_n.

3.3.4 Program. A program for finding a minimal word equivalent to $w \in F_n$.

```
MinWord:=function(w,G)
#  This  function  takes  a  word  w  in  F_n,  and  a  free  group  G=F_n.
#  It  returns  w  if  w  is  minimal,  and  returns  a  minimal  word  equivalent
#  to  w  if  w  is  not  minimal.
local  wa2,  b;
wa2:=WA2(G);
#  Minimality  Test
if  IsMin(w,G)=true  then
     return  w;
else
   repeat
     for  b  in  wa2  do
          if  Length(Image(b,w))<Length(w)  then
               w:=Image(b,w);
          fi;
     od;
   until  IsMin(w,G)=true;
   return  w;
fi;
end;
```

Given a word $w \in F_n$, we can represent w by a vector. We refer to such vectors as "External representatives of that word". In our preferred programming language GAP, we view such vector as follows:

3.3.5 Definition. The **External representative** (ext rep for short) of a word $w = x_1^{\delta_1} x_2^{\delta_2} \cdots x_l^{\delta_l}$ of length l is defined as $w = [1, \delta_1, 2, \delta_2, \cdots, l, \delta_l]$ where $\delta_i \in \{-1, 1\}$ for $1 \le i \le l$.

3.3.6 Example. In $F_2 = \langle x_1, x_2 \rangle$, the external representatives of $x_2 x_1^{-1}$ and $x_1 x_2 x_1^{-1} x_2^{-1}$ are $[2, 1, 1, -1]$ and $[1, 1, 2, 1, 1, -1, 2, -1]$ respectively.

We shall sometimes represent input and output of our program as external representatives. Whenever this is done, it is an easy task to convert back to the coreesponding word.

3.3.7 Program. A program for finding a list of all minimal words equivalent to $w \in F_n$.

```
Min_Eq_Elts:=function(w,G)
# This function asks for G=F_n, and an external representative w.
# It returns the list of all minimal words equivalent to w in F_n.
local wa1,wa2,MinEquivElts,a,b,v,wg,j,Ans;
wa1:=WA1(G); # This is the list of Type 1 Whitehead Automorphisms.
wa2:=WA2(G); # This is the list of Type 2 Whitehead Automorphisms.
MinEquivElts:=[];
# Coverting w to a word in G.
v:=GeneratorsOfGroup(G);
wg:=G.1^0;
for j in [1..(Length(w)/2)] do
    wg:=wg*v[w[2*j-1]]^w[2*j];
od;
w:=wg; # This is the corresponding word in G.
Add(MinEquivElts,MinWord(w,G)); w:=MinEquivElts[1];
# We now append equivalent minimal words.
for w in MinEquivElts do
    for a in wa1 do if Length(Image(a,w))=Length(w) then
        AddSet(MinEquivElts,Image(a,w));
    fi; od;
    for b in wa2 do if Length(Image(b,w))=Length(w) then
        AddSet(MinEquivElts,Image(b,w));
    fi; od;
od;
Ans:=List(MinEquivElts,x->ExtRepOfObj(x));
return Ans;
end;
```

3.3.8 Program. Another program for finding a list of all minimal words equivalent to $w \in F_n$.

```
MinEqElts:=function(w,G)
# This function takes a word w in F_n, and a free group G=F_n.
# It returns the list of all minimal words equivalent to w in F_n.
local wa1,wa2,MinEquivElts,a,b;
wa1:=WA1(G); # This is the list of Type 1 Whitehead Automorphisms.
wa2:=WA2(G); # This is the list of Type 2 Whitehead Automorphisms.
```

```
MinEquivElts:=[];  Add(MinEquivElts,MinWord(w,G));
w:=MinEquivElts[1];
# We now append equivalent minimal words.
for w in MinEquivElts do
    for a in wa1 do if Length(Image(a,w))=Length(w) then
        AddSet(MinEquivElts,Image(a,w));
    fi; od;
    for b in wa2 do if Length(Image(b,w))=Length(w) then
        AddSet(MinEquivElts,Image(b,w));
    fi; od;
od;
return MinEquivElts;
end;
```

3.3.9 Program. A program for checking whether two words in F_n are equivalent.

```
IsEquiv:=function(w,w1,G)
# IsEquiv asks for G=F_n, and external representatives w and
# w1 of two words in G. It returns true if the corresponding
# words are equivalent, and false if otherwise.
local v,wg,w1g,j;
v:=GeneratorsOfGroup(G);
wg:=G.1^0;  w1g:=wg;
# Converting w to its corresponding word wg.
for j in [1..(Length(w)/2)] do
    wg:=wg*v[w[2*j−1]]^w[2*j];
od;
# Converting w1 to its corresponding word w1g.
for j in [1..(Length(w1)/2)] do
    w1g:=w1g*v[w1[2*j−1]]^w1[2*j];
od;
w:=wg;  w1:=w1g;
if (MinWord(w,G) in MinEqElts(w1,G))=true then
    return true;
else
    return false;
fi;
end;
```

3.3.10 Program. A program for classifying all minimal words of length l in F_n up to equivalence.

```
EquivCl:=function(n,l)
# EquivCl takes a word length l, and the number of generators n of F_n.
# It finds all minimal words of such length in F_n, classifies them
# up to equivalence, and returns the result.
local MW,li,w,A,G,found,MWE;
G:=FreeGroup(n);  MW:=MinimalWords(n,l);
MWE:=List(MW,x−>ExtRepOfObj(x));
li:=[];  # Empty list of equivalent minimal words.
```

```
for w in MWE do
    found:=false;
    for A in li do
        if (w in A)=true then
            found:=true;
            break;
        elif (IsEquiv(w,A[1],G))=true then
            Append(A, Min_Eq_Elts(w,G));
            found:=true;
            break;
        fi;
    od;
    if found=false then
        Add(li , Min_Eq_Elts(w,G));
    fi;
od;
return li;
end;
```

We now introduce terms that will enable us classify minimal words of length l in F_n up to equivalence.

3.3.11 Definition.   • (T1) A **cyclic structure** of a reduced word w is a cyclic representation of w.

   • (T2) Two non-identity elements of a free group are said to be in the **same cyclic structure** (say $\langle w \rangle$) if both words represent the same cyclic word.

   • (T3) A reduced word is called **exact** if no other reduced word can be in its cyclic structure. In other words, if it is the only reduced word obtainable from its cyclic structure.

3.3.12 Example.   • In $F_2 = \langle x, y \rangle$, xxy, xyx and yxx are in the same cyclic structure whereas xyx and xyy are not in the same cyclic structure.

   • xyx is not exact (since $|\langle xyx \rangle| = |\{xxy, xyx, yxx\}| = 3 > 1$). On the other hand, xxx is exact.

Below are some immediate consequences.

3.3.13 Proposition. Every exact word is minimal.

The converse of Proposition 3.3.13 is not necessarily true as $xy\bar{x}\bar{y}$ is minimal, but not exact. A natural question is:

3.3.14 Question. Does the converse of Proposition 3.3.13 hold for any word length?

We answer as follows:

3.3.15 Lemma.   • (U1) Every minimal word of length 2 or 3 is exact.

   • (U2) There is no such full characterization for minimal words of length $l \geq 4$ in $F_{n \geq 2}$.

3.3.16 Corollary. There is only one equivalence class of minimal words of lengths 2 and 3 in $F_{n \geq 2}$. Furthermore, the number of elements in such equivalence class is $2n$.

3.3.17 Notation. The minimal exact words will be referred to as the **trivial minimal words** while the minimal non-exact words will be called the **non-trivial minimal words**.

Next, we define our preferred ordering.

3.3.18 Definition. Let $L_n = \{f_1, f_1^{-1}, f_2, f_2^{-1}, \cdots, f_n, f_n^{-1}\}$. Given a well-ordering $\leq$ on L_n, we define a **well-ordering on** L_n^* as follows: if $a = a_1 a_2 \cdots a_l$ and $b = b_1 b_2 \cdots b_m$, then $a < b$ if and only if either $l < m$ or $l = m$, and $a_j = b_j$ for $j \leq i \leq l$ (with $a_{i+1} < b_{i+1}$).

3.3.19 Example. Take $L_2 = \{f_1, f_1^{-1}, f_2, f_2^{-1}\}$ with the ordering $f_1^{-1} < f_2^{-1} < f_1 < f_2$. We view the irreducible words of length 2 in F_2 as $f_1^{-2} < f_2^{-2} < f_1^2 < f_2^2$.

3.3.20 Remark. In the sequel, we shall take the representative of a class of equivalent minimal words to be the least element in that class with respect to the ordering defined in Definition 3.3.18. From our program, that representative is the first element in the sublist under consideration. For example, consider the minimal words of length 3 in F_3 w.r.t the ordering described above. By Corollary 3.3.16, there is only one such class. So, we take the representative to be "f_1^{-3}".

Let n denote the rank of a free group, l word length, M the number of minimal words of length l in F_n, N is the number of (distinct) equivalence classes of minimal words of length l in F_n, and Card the respective cardinality of each equivalence class. We give a summary of our results as follows:

n	l	M	N	Class representatives	Card
2	2	4	1	f_1^{-2}	4
2	3	4	1	f_1^{-3}	4
2	4	44	3	$f_1^{-4}, f_1^{-2}f_2^{-2}, f_1^{-1}f_2^{-1}f_1f_2$	$4, 32, 8$
2	5	164	4	$f_1^{-5}, f_1^{-3}f_2^{-2}, f_1^{-2}f_2^{-1}f_1^{-1}f_2, f_1^{-2}f_2^{-1}f_1f_2$	$4, 80, 40, 40$
2	6	436	10	$f_1^{-6}, f_1^{-4}f_2^{-2}, f_1^{-3}f_2^{-1}f_1^{-1}f_2, f_1^{-3}f_2^{-1}f_1f_2,$ $f_1^{-3}f_2^{-3}, f_1^{-2}f_2^{-1}f_1^{-1}f_2^{-1}f_2, f_1^{-2}f_2^{-1}f_1^{-1}f_2^2,$ $f_1^{-2}f_2^{-1}f_1^2f_2, f_1^{-2}f_2^{-2}f_1^{-1}f_2, f_1^{-2}f_2^{-2}f_1f_2$	$4, 120, 48, 48, 24, 24, 48,$ $48, 48, 24$
3	2	6	1	f_1^{-2}	6
3	3	6	1	f_1^{-3}	6
3	4	126	3	$f_1^{-4}, f_1^{-2}f_2^{-2}, f_1^{-1}f_2^{-1}f_1f_2$	$6, 96, 24$
3	5	486	4	$f_1^{-5}, f_1^{-3}f_2^{-2}, f_1^{-2}f_2^{-1}f_1^{-1}f_2, f_1^{-2}f_2^{-1}f_1f_2$	$6, 240, 120, 120$
3	6	3270	11	$f_1^{-6}, f_1^{-4}f_2^{-2}, f_1^{-3}f_2^{-1}f_1^{-1}f_2, f_1^{-3}f_2^{-1}f_1f_2,$ $f_1^{-3}f_2^{-3}, f_1^{-2}f_2^{-1}f_1^{-1}f_2, f_1^{-2}f_2^{-1}f_1^2f_2,$ $f_1^{-2}f_2^{-2}f_1^{-1}f_2, f_1^{-2}f_2^{-2}f_1f_2, f_1^{-2}f_2^{-2}f_3^{-2},$ $f_1^{-2}f_2^{-1}f_1^2f_2$	$6, 360, 144, 144, 72, 72,$ $144, 144, 144, 1968, 72$
4	2	8	1	f_1^{-2}	8
4	3	8	1	f_1^{-3}	8
4	4	248	3	$f_1^{-4}, f_1^{-2}f_2^{-2}, f_1^{-1}f_2^{-1}f_1f_2$	$8, 192, 48$

n	l	M	N	Class representatives	Card
4	5	968	4	$f_1^{-5}, f_1^{-3}f_2^{-2}, f_1^{-2}f_2^{-1}f_1^{-1}f_2, f_1^{-2}f_2^{-1}f_1f_2$	$8, 480, 240, 240$
4	6	10472			
5	2	10	1	f_1^{-2}	10
5	3	10	1	f_1^{-3}	10
5	4	410	3	$f_1^{-4}, f_1^{-2}f_2^{-2}, f_1^{-1}f_2^{-1}f_1f_2$	$10, 320, 80$
5	5	1610	4	$f_1^{-5}, f_1^{-3}f_2^{-2}, f_1^{-2}f_2^{-1}f_1^{-1}f_2, f_1^{-2}f_2^{-1}f_1f_2$	$10, 800, 400, 400$
5	6	24010			
6	2	12	1	f_1^{-2}	12
6	3	12	1	f_1^{-3}	12
6	4	612	3	$f_1^{-4}, f_1^{-2}f_2^{-2}, f_1^{-1}f_2^{-1}f_1f_2$	$12, 480, 120$
6	5	2412	4	$f_1^{-5}, f_1^{-3}f_2^{-2}, f_1^{-2}f_2^{-1}f_1^{-1}f_2, f_1^{-2}f_2^{-1}f_1f_2$	$12, 1200, 600, 600$

See Appendix B for some results of this classifcation.

4. Concluding Remarks

4.1 Summary

In Chapter one, we visited some basic properties of free groups as well as Nielsen's approach to the Subgroup theorem. We began Chapter two with a brief survey of automorphisms of free groups, starting with what we termed "Special automorphisms" to Nielsen automorphisms, and lastly Whitehead automorphisms. We denoted these sets of automorphisms by ψ_n, ω_n and Ω_n respectively; and remarked that $\psi_n \subset \omega_n \subset \Omega_n$. In Section 2.4, we studied the Whitehead's automorphisms in two forms: "Type 1 Whitehead automorphisms" and "Type 2 Whitehead automorphisms". We asserted in Remark 2.4.9 that Type 1 Whitehead automorphisms are just permutations, and so do not reduce the length of a word while Type 2 Whitehead automorphisms may reduce the length of a word. We went further in Notation 2.4.19 and Remark 2.4.20 to subdivide the Type 2 Whitehead automorphisms into three types, namely the "Bad", "Maybe Good" and "Good" Type 2 Whitehead automorphisms, and characterised all good type 2 Whitehead automorphisms for non-minimal words of length 2 in F_2. We also noted that it is not easy in theory to characterise all the good type 2 Whitehead automorphisms for a non-minimal word of lengths $l \geq 4$ in free groups of finite rank $n \geq 2$. However, this is a trivial problem in programming.

We also studied automorphisms of the Abelianization of F_n, with special consideration to $Aut(\bar{F}_2)$. Furthermore, we introduced the backbone of this study "The Whitehead's theorem"; and then used operations with "Character pairs" to investigate equivalence of elements under $Aut(F_n)$ in Section 2.5 following [HL74] and [LS77]. We studied a presentation of $Aut(F_n)$ given by McCool, and conclude our discussions in chapter two with an insight to other presentations of $Aut(F_n)$ in Remark 2.6.15.

In Chapter three, we revisited the Whitehead's theorem, and gave a brief remark on the need for Whitehead Algorithm. We went further in Lemma 3.1.3 to ascertain that two irreducible(minimal) words of different lengths are not equivalent. This formed the basis of the introduction of Whitehead's algorithm in Section 3.2. Moreover, we used [GAP13] in Section 3.3 to design programs...ranging from programs that can be used to appreciate Types 1 and 2 Whitehead automorphisms to programs that can be used to determine whether two elements of a finitely generated free group are equivalent or not. Finally, we classified all minimal words of lengths $2, 3, 4, 5$ and 6 in F_n (for some $n \in [2, 6]$) up to equivalence.

4.2 Further Research Questions

Let M_{ln} denote the set of all Minimal words of length l in $F_{n \geq 2}$. From our discussions in Section 3.3, we deduce that:

- $|M_{2n}| = 2n = |M_{3n}|$;

- There is only one equivalence class of trivial minimal words. The representatives can be seen as f_1^{-2} and f_1^{-3} for $l = 2$ and $l = 3$ respectively.

4.2.1 Question. • (V1) Is it possible to characterize all representatives of equivalence classes of minimal words of lengths $l \geq 4$ in $F_{n \geq 2}$?

- (V2) What can be said about $|M_{ln}|$ for $l \geq 4$?

We answer (V1) and (V2) for $l = 4$ and $l = 5$ as follows:

4.2.2 Conjecture.   • (W1) There are exactly three equivalence classes of minimal words of length 4 in $F_{n \geq 2}$. Their representatives can be seen as $f_1^{-4}, f_1^{-1}f_2^{-1}f_1f_2$ and $f_1^{-2}f_2^{-2}$ with cardinalities $2n$, $4n(n-1)$ and $16n(n-1)$ respectively.

• (W2) $|M_{4n}| = 2n(10n - 9)$.

4.2.3 Conjecture.   • (X1) There are exactly four equivalence classes of minimal words of length 5 in $F_{n \geq 2}$. Their representatives can be seen as f_1^{-5}, $f_1^{-3}f_2^{-2}$, $f_1^{-2}f_2^{-1}f_1^{-1}f_2$ and $f_1^{-2}f_2^{-1}f_1f_2$ with cardinalities $2n$, $40n(n-1)$, $20n(n-1)$ and $20n(n-1)$ respectively.

• (X2) $|M_{5n}| = 2n(40n - 39)$.

One possible future work is to prove our Conjectures 4.2.2 and 4.2.3 above as well as answer (V1) and (V2) for $l \geq 6$. Another important observation is that only f_1 and f_2 are enough to describe all the representatives of equivalent classes of minimal words of lengths 2, 3, 4 and 5 in $F_{n \geq 2}$. On the other hand, the two letters are not sufficient to describe all the representatives of equivalence classes of minimal words of length 6. Hence, it is important to investigate the following:

4.2.4 Question. How many letters do we need to describe all the representatives of equivalence classes of minimal words of any given length?

4.3   Conclusions

The Whitehead algorithm is a powerful tool for classifying all minimal words of a certain length l in a finitely generated free group F_n up to equivalence. We introduced the algorithm in Section 3.2, and used [GAP13] to give the resulting programs in Section 3.3.

Furthermore, we classified all minimal words of lengths l for $2 \leq l \leq 6$ and some $n \in [2, 6]$ up to equivalence, and conclude by investigating the nature of the resulting representatives. We look forward to extending this in future works.

Appendix A. All reduced words and minimal words of lengths l in F_n

A.1 A program for finding all reduced words of lengths l in F_n

```
ReducedWords:=function(n,l)
# ReducedWords takes the number of generators of F_n, and word
# length l. It returns all reduced words of such length in F_n.
local G,gens,i,j,li,lis,T,p;
G:=FreeGroup(n); gens:=GeneratorsOfGroup(G); li:=[]; lis:=[];
for i in [1..n] do for j in [−1,1] do
    Add(li,gens[i]^j);
od; od;
T:=Tuples(li,l); # T is the set of all ordered tuples of length l.
for p in [1..Length(T)] do
    if Length(Product(T[p]))=l then
        Add(lis,Product(T[p]));
    fi;
od;
return lis;
end;
```

A.2 A program for finding all minimal words of lengths l in F_n

```
MinimalWords:=function(n,l)
# MinimalWords takes the number of generators of F_n, and word
# length l. It returns all minimal words of such length in F_n.
local li, w, x, j, v, wg, G, RW, RWE;
G:=FreeGroup(n); li:=[]; RW:=ReducedWords(n,l);
# Obtain list of external representatives of words in RW.
RWE:=List(RW,x−>ExtRepOfObj(x)); # The desired list of ext reps.
# Append minimal elements from RW to the list li.
for w in RWE do
    # Coverting an element of RWE to a word in G.
    v:=GeneratorsOfGroup(G); wg:=G.1^0;
    for j in [1..(Length(w)/2)] do
        wg:=wg*v[w[2*j−1]]^w[2*j];
    od;
    # Minimality test
    if IsMin(wg,G)=true then
        Add(li,wg);
    fi;
od;
return li;
end;
```

Appendix B. Equivalence classes of minimal words of certain lengths in F_2 and F_3

B.1   Equivalence classes of minimal words of lengths $2, 3, 4, 5$ and 6 in F_2

EquivCl(2,2):=[[f1^-2, f2^-2, f1^2, f2^2]].

EquivCl(2,3):=[[f1^-3, f2^-3, f1^3, f2^3]].

EquivCl(2,4):=[[f1^-4, f2^-4, f1^4, f2^4],
[f1^-2*f2^-2, f1^-1*f2^-2*f1^-1, f2^-1*f1^-2*f2^-1, f2^-2*f1^-2,
f1^-1*f2^-1*f1^-1*f2, f2^-1*f1^-1*f2^-1*f1, f1^-1*f2^-1*f1*f2^-1,
f2^-1*f1^-1*f2*f1^-1, f1^-2*f2^2, f2^-2*f1^2, f1^-1*f2*f1^-1*f2^-1,
f2^-1*f1*f2^-1*f1^-1, f1^-1*f2^2*f1^-1, f2^-1*f1^2*f2^-1,
f1^-1*f2*f1*f2, f2^-1*f1*f2*f1, f1*f2^-1*f1^-1*f2^-1,
f2*f1^-1*f2^-1*f1^-1, f1*f2^-2*f1, f2*f1^-2*f2, f1*f2^-1*f1*f2,
f2*f1^-1*f2*f1, f1^2*f2^-2, f2^2*f1^-2, f1*f2*f1^-1*f2, f2*f1*f2^-1*f1,
 f1*f2*f1*f2^-1, f2*f1*f2*f1^-1, f1^2*f2^2, f1*f2^2*f1,
f2*f1^2*f2, f2^2*f1^2],
[f1^-1*f2^-1*f1*f2, f2^-1*f1^-1*f2*f1, f1^-1*f2*f1*f2^-1,
f2^-1*f1*f2*f1^-1, f1*f2^-1*f1^-1*f2, f2*f1^-1*f2^-1*f1,
f1*f2*f1^-1*f2^-1, f2*f1*f2^-1*f1^-1]].

EquivCl(2,5):=[[f1^-5, f2^-5, f1^5, f2^5],
[f1^-3*f2^-2, f1^-2*f2^-2*f1^-1, f1^-2*f2^-3, f1^-1*f2^-2*f1^-2,
f1^-1*f2^-2*f1^-1*f2, f1^-1*f2^-3*f1^-1, f2^-1*f1^-3*f2^-1,
f2^-1*f1^-2*f2^-2*f1^-1, f2^-1*f1^-2*f2^-2, f2^-2*f1^-3,
f2^-2*f1^-2*f2^-1, f2^-3*f1^-2, f1^-3*f2^2, f1^-2*f2^-1*f1*f2^-1,
f1^-1*f2^-1*f1^-1*f2^2, f2^-1*f1^-1*f2^-1*f1^2, f2^-2*f1^-1*f2*f1^-1,
f2^-3*f1^2, f1^-1*f2^-1*f1*f2^-1*f1^-1, f2^-1*f1^-1*f2*f1^-1*f2^-1,
f1^-2*f2*f1*f2, f1^-2*f2^2*f1^-1, f1^-2*f2^3, f1^-1*f2^-1*f1^2*f2^-2,
f2^-1*f1^-1*f2^2*f1^-1, f2^-2*f1^3, f2^-2*f1^2*f2^-1, f2^-2*f1*f2*f1,
f1^-1*f2*f1^-1*f2^-2, f2^-1*f1*f2^-1*f1^-2, f1^-1*f2^2*f1^-2,
f1^-1*f2^2*f1^-1*f2^-1, f2^-1*f1^2*f2^-1*f1^-1, f2^-1*f1^2*f2^-2,
f1^-1*f2*f1^2*f2, f1^-1*f2*f1*f2*f1^-1, f1^-1*f2^3*f1^-1,
f2^-1*f1^3*f2^-1, f2^-1*f1*f2*f1*f2^-1, f2^-1*f1*f2^2*f1,
f1*f2^-1*f1^-2*f2^-1, f1*f2^-1*f1^-1*f2^-1*f1, f1*f2^-3*f1,
f2*f1^-3*f2, f2*f1^-1*f2^-1*f1^-1*f2, f2*f1^-1*f2^-2*f1^-1,
f1*f2^-2*f1^2, f1*f2^-2*f1*f2, f2*f1^-2*f2*f1, f2*f1^-2*f2^2,
f1*f2^-1*f1*f2^2, f2*f1^-1*f2*f1^2, f1^2*f2^-1*f1^-1*f2^-1,
f1^2*f2^-2*f1, f1^2*f2^-3, f1*f2*f1^-2*f2,
f2*f1*f2^-2*f1, f2^2*f1^-3, f2^2*f1^-2*f2^-1*f1^-1,
f1*f2*f1^-1*f2*f1, f2*f1*f2^-1*f1*f2, f1^3*f2^-2, f1^2*f2*f1^-1*f2,
f1*f2*f1*f2^-2, f2*f1*f2*f1^-2, f2^2*f1*f2^-1*f1, f2^3*f1^-2, f1^3*f2^2,

```
f1^2*f2^2*f1 ,  f1^2*f2^3,  f1*f2^2*f1^2,
f1*f2^2*f1*f2^-1,  f1*f2^3*f1,  f2*f1^3*f2 ,  f2*f1^2*f2*f1^-1,
f2*f1^2*f2^2,  f2^2*f1^3,  f2^2*f1^2*f2,  f2^3*f1^2 ],
[ f1^-2*f2^-1*f1^-1*f2 ,  f1^-1*f2^-1*f1^-2*f2 ,  f2^-1*f1^-1*f2^-2*f1 ,
f2^-2*f1^-1*f2^-1*f1 ,  f1^-1*f2^-1*f1^-1*f2*f1^-1,  f1^-1*f2^-2*f1*f2^-1,
f2^-1*f1^-2*f2*f1^-1,  f2^-1*f1^-1*f2^-1*f1*f2^-1,
f1^-2*f2*f1^-1*f2^-1,  f1^-1*f2^-1*f1*f2^-2,  f2^-1*f1^-1*f2*f1^-2,
f2^-2*f1*f2^-1*f1^-1,  f1^-1*f2*f1^-2*f2^-1,  f1^-1*f2*f1^-1*f2^-1*f1^-1,
f2^-1*f1*f2^-1*f1^-1*f2^-1,  f2^-1*f1*f2^-2*f1^-1,
f1^-1*f2*f1*f2^2,  f1^-1*f2^2*f1*f2,  f2^-1*f1^2*f2*f1,  f2^-1*f1*f2*f1^2,
f1*f2^-1*f1^-1*f2^-2,  f1*f2^-2*f1^-1*f2^-1,  f2*f1^-2*f2^-1*f1^-1,
f2*f1^-1*f2^-1*f1^-2,  f1*f2^-1*f1^2*f2 ,
f1*f2^-1*f1*f2*f1,  f2*f1^-1*f2*f1*f2,  f2*f1^-1*f2^2*f1 ,
f1^2*f2^2-1*f1*f2,  f1*f2*f1^-1*f2^2,  f2*f1*f2^-1*f1^2,  f2^2*f1^-1*f2*f1,
f1*f2*f1*f2^-1*f1,  f1*f2^2*f1^-1*f2,  f2*f1^2*f2^-1*f1,  f2*f1*f2*f1^-1*f2 ,
f1^2*f2*f1*f2^-1,  f1*f2*f1^2*f2^-1,  f2*f1*f2^2*f1^-1,  f2^2*f1*f2*f1^-1 ],
[ f1^-2*f2^-1*f1*f2,  f1^-1*f2^-2*f1*f2,  f2^-1*f1^-1*f2*f1 ,
f2^-2*f1^-1*f2*f1,  f1^-2*f2*f1*f2^-1,  f1^-1*f2^-1*f1^2*f2,
f2^-1*f1^-1*f2*f1*f1,  f1^-2*f2*f1*f2^-1,  f1^-1*f2^-1*f1*f2^2,
f1^-1*f2^-1*f1*f2*f1^-1,  f1^-1*f2^-1*f1*f2^2,  f2^-1*f1^-1*f2*f1^2,
f2^-1*f1^-1*f2*f1*f2^-1,  f2^-1*f1^-1*f2^2*f1,  f2^-2*f1*f2*f1^-1,
f1^-1*f2*f1*f2^-1*f1^-1,  f1^-1*f2*f1*f2^-2,  f2^-1*f1*f2*f1^-2,
f2^-1*f1*f2*f1^-1*f2^-1,  f1^-1*f2*f1^2*f2^-1,  f1^-1*f2^2*f1*f2^-1,
f2^-1*f1^2*f2*f1^-1,  f2^-1*f1*f2^2*f1^-1,  f1*f2^-1*f1^-2*f2 ,
f1*f2^-2*f1^-1*f2,  f2*f1^-2*f2^-1*f1,  f2*f1^-1*f2^-2*f1 ,
f1*f2^-1*f1^-1*f2*f1,  f1*f2^-1*f1^-1*f2^2,  f2*f1^-1*f2^-1*f1^2,
f2*f1^-1*f2^-1*f1*f2,  f1^2*f2^-1*f1^-1*f2,  f1*f2*f1^-2*f2^-1,
f1*f2*f1^-1*f2^-1*f1,  f1*f2*f1^-1*f2^-2,  f2*f1*f2^-1*f1^-2,
f2*f1*f2*f1^-1*f2^-1,  f2*f1*f2^-2*f1^-1,  f2^2*f1^-1*f2^-1*f1 ,
f1^2*f2*f1^-1*f2^-1,  f1*f2^2*f1^-1*f2^-1,  f2*f1^2*f2^-1*f1^-1,
f2^2*f1*f2^-1*f1^-1 ] ].

EquivCl(2,6):=[ [ f1^-6,  f2^-6,  f1^6,  f2^6 ],
[ f1^-4*f2^-2,  f1^-4*f2^2,  f1^-3*f2^-1*f1*f2^-1,  f1^-3*f2^-2*f1^-1,
f1^-2*f2^-2*f1^-2,  f1^-2*f2^-4,  f1^-1*f2^-2*f1^-3,
f1^-1*f2^-2*f1^-1*f2^2,  f1^-1*f2^-3*f1^-1*f2 ,  f1^-1*f2^-4*f1^-1,
f2^-1*f1^-4*f2^-1,  f2^-1*f1^-3*f2^-1*f1,  f2^-1*f1^-2*f2^-1*f1^2,
f2^-1*f1^-2*f2^-3,  f2^-2*f1^-4,  f2^-2*f1^-2*f2^-2,  f2^-3*f1^-2*f2^-1,
f2^-3*f1^-1*f2*f1^-1,  f2^-4*f1^-2,  f2^-4*f1^2,  f1^-3*f2*f1*f2 ,
f1^-3*f2^2*f1^-1,  f1^-2*f2^-1*f1^2*f2^-1,  f1^-2*f2^-1*f1*f2^-1*f1^-1,
f1^-1*f2^-1*f1^-1*f2^3,  f2^-1*f1^-1*f2^-1*f1^3,
f2^-2*f1^-1*f2*f1^-1*f2^-1,  f2^-2*f1^-1*f2^2*f1^-1,  f2^-3*f1^2*f2^-1,
f2^-3*f1*f2*f1,  f1^-1*f2^-1*f1*f2^-1*f1^-2,  f2^-1*f1^-1*f2*f1^-1*f2^-2,
f1^-2*f2*f1^2*f2,  f1^-2*f2*f1*f2*f1^-1,  f1^-2*f2^2*f1^-2,  f1^-2*f2^4,
f1^-1*f2^-1*f1^3*f2^-1,  f1^-1*f2^-1*f1^2*f2^-1*f1^-1,
f2^-1*f1^-1*f2^2*f1^-1*f2^-1,  f2^-1*f1^-1*f2^3*f1^-1,  f2^-2*f1^4,
f2^-2*f1^2*f2^-2,  f2^-2*f1*f2*f1*f2^-1,  f2^-2*f1*f2^2*f1 ,
f1^-1*f2*f1^-1*f2^-3,  f2^-1*f1*f2^-1*f1^-3,  f1^-1*f2^2*f1^-3,
f1^-1*f2^2*f1^-1*f2^-2,  f2^-1*f1^2*f2^-1*f1^-2,  f2^-1*f1^2*f2^-3,
```

f1^-1*f2*f1^3*f2 , f1^-1*f2*f1^2*f2*f1^-1, f1^-1*f2*f1*f2*f1^-2,
f1^-1*f2^3*f1^-1*f2^-1, f1^-1*f2^4*f1^-1, f2^-1*f1^4*f2^-1,
f2^-1*f1^3*f2^-1*f1^-1, f2^-1*f1*f2*f1*f2^-2, f2^-1*f1*f2^2*f1*f2^-1,
f2^-1*f1*f2^3*f1, f1*f2^-1*f1^-3*f2^-1, f1*f2^-1*f1^-2*f2^-1*f1,
f1*f2^-1*f1^-1*f2^-1*f1^2, f1*f2^-3*f1*f2, f1*f2^-4*f1, f2*f1^-4*f2,
f2*f1^-3*f2*f1, f2*f1^-1*f2^-1*f1^-1*f2^2, f2*f1^-1*f2^-2*f1^-1*f2,
f2*f1^-1*f2^-3*f1^-1, f1*f2^-2*f1*f2^2, f1*f2^-2*f2*f1^2,
f2*f1^-2*f2^3, f1*f2^-1*f1*f2^3, f2*f1^-1*f2*f1^3,
f1^2*f2^-1*f1^-2*f2^-1, f1^2*f2^-1*f1^-1*f2^-1*f1, f1^2*f2^-2*f1^2,
f1^2*f2^-4, f1*f2*f1^-3*f2, f1*f2*f1^-2*f2*f1, f2*f1*f2^-2*f1*f2,
f2*f1*f2^-3*f1, f2^2*f1^-4, f2^2*f1^-2*f2^2, f2^2*f1^-1*f2^-1*f1^-1*f2,
f2^2*f1^-1*f2^-2*f1^-1, f1*f2*f1^-1*f2*f1^2, f2*f1*f2^-1*f1*f2^2,
f1^3*f2^-1*f1^-1*f2^-1, f1^3*f2^-2*f1, f1^2*f2*f1^-2*f2,
f1^2*f2*f1^-1*f2*f1, f1*f2*f1*f2^-3, f2*f1*f2*f1^-3,
f2^2*f1*f2^-1*f1*f2, f2^2*f1*f2^-2*f1, f2^3*f1^-2*f2,
f2^3*f1^-1*f2^-1*f1^-1, f1^4*f2^-2, f1^4*f2^2, f1^3*f2*f1^-1*f2,
f1^3*f2^2*f1, f1^2*f2^2*f1^2, f1^2*f2^4, f1*f2^2*f1^3, f1*f2^2*f1*f2^-2,
f1*f2^3*f1*f2^-1, f1*f2^4*f1, f2*f1^3*f2*f1^-1,
f2*f1^2*f2*f1^-2, f2*f1*f2^3, f2^2*f1^4, f2^2*f1^2*f2^2,
f2^3*f1^2*f2, f2^3*f1*f2^-1*f1, f2^4*f1^-2, f2^4*f1^2],
[f1^-3*f2^-1*f1^-1*f2, f1^-2*f2^-1*f1^-1*f2*f1^-1, f1^-1*f2^-1*f1^-3*f2 ,
f1^-1*f2^-3*f1*f2^-1, f2^-1*f1^-3*f2*f1^-1, f2^-1*f1^-1*f2^-3*f1,
f2^-2*f1^-1*f2^-1*f1*f2^-1, f2^-3*f1^-1*f2^-1*f1, f1^-3*f2*f1^-1*f2^-1,
f1^-1*f2^-1*f1^-1*f2*f1^-2, f2^-1*f1^-1*f2^-1*f1*f2^-2,
f2^-3*f1*f2^-1*f1^-1, f1^-2*f2*f1^-1*f2^-1*f1^-1, f1^-1*f2^-1*f1*f2^-3,
f2^-1*f1^-1*f2*f1^-3, f2^-2*f1*f2^-1*f1^-1*f2^-1, f1^-1*f2*f1^-3*f2^-1,
f1^-1*f2*f1^-1*f2^-1*f1^-2, f2^-1*f1*f2^-1*f1^-1*f2^-2,
f2^-1*f1*f2^-3*f1^-1, f1^-1*f2*f1^2*f2^3, f1*f2^-1*f1^-1*f2^-3,
f1*f2^-3*f1^-1*f2^-1, f2*f1^-3*f2^-1*f1^-1, f2*f1^-1*f2^-1*f1^-1*f1^-3,
f1*f2^-1*f1^3*f2, f1*f2^-1*f1*f2*f1^2, f2*f1^-1*f2*f1*f2^2,
f2*f1^-1*f2^3*f1, f1^2*f2^-1*f1*f2*f1, f1*f2*f1^-1*f2^3,
f2*f1*f2^-1*f1^3, f2^2*f1^-1*f2*f1*f2, f1^3*f2^-1*f1*f2,
f1*f2*f1*f2^-1*f1^2, f2*f1*f2*f1^-1*f2^2, f2^3*f1^-1*f2*f1,
f1^3*f2*f1*f2^-1, f1^2*f2*f1*f2^-1*f1, f1*f2*f1^3*f2^-1,
f1*f2^3*f1^-1*f2, f2*f1^3*f2^-1*f1, f2*f1*f2^3*f1^-1,
f2^2*f1*f2*f1^-1*f2, f2^3*f1*f2*f1^-1],
[f1^-3*f2^-1*f1*f2, f1^-1*f2^-3*f1*f2, f2^-1*f1^-3*f2*f1,
f2^-3*f1^-1*f2*f1, f1^-3*f2*f1*f2^-1, f1^-2*f2^-1*f1*f2*f1^-1,
f2^-2*f1^-1*f2*f1*f2^-1, f2^-3*f2*f1*f2^-1, f1^-2*f2*f1*f2^-1*f1^-1,
f1^-1*f2^-1*f1^3*f2, f1^-1*f2^-1*f1*f1*f1^-2, f1^-1*f2^-1*f1*f2^3,
f2^-1*f1^-1*f2*f1^3, f2^-1*f1^-1*f2*f1*f2^-2, f2^-1*f1^-1*f2^3*f1,
f2^-2*f1*f2*f1^-1*f2^-1, f1^-1*f2*f1*f2^-1*f1^-2, f1^-1*f2*f1*f2^-3,
f2^-1*f1*f2*f1^-3, f2^-1*f1*f2*f1^-1*f2^-2, f1^-1*f2*f1^3*f2^-1,
f1^-1*f2^3*f1*f2^-1, f2^-1*f1^3*f2*f1^-1, f2^-1*f1*f2^3*f1^-1,
f1*f2^-1*f1^-3*f2, f1*f2^-3*f1^-1*f2, f2*f1^-3*f2^-1*f1,
f2*f1^-1*f2^-3*f1, f1*f2^-1*f1^-1*f2*f1^2, f1*f2^-1*f1^-1*f2^3,
f2*f1^-1*f2^-1*f1^3, f2*f1^-1*f2^-1*f1*f1*f2^2, f1^2*f2^-1*f1^-1*f2*f1,

f1*f2*f1^-3*f2^-1, f1*f2*f1^-1*f2^-1*f1^2, f1*f2*f1^-1*f2^-3,
f2*f1*f2^-1*f1^-3, f2*f1*f2^-1*f1^-1*f2^2, f2*f1*f2^-3*f1^-1,
f2^2*f1^-1*f2^-1*f1*f2, f1^3*f2^-1*f1^-1*f2, f1^2*f2*f1^-1*f2^-1*f1,
f2^2*f1*f2^-1*f1^-1*f2, f2^3*f1^-1*f2^-1*f1, f1^3*f2*f1^-1*f2^-1,
f1*f2^3*f1^-1*f2^-1, f2*f1^3*f2^-1*f1^-1, f2^3*f1*f2^-1*f1^-1],
[f1^-3*f2^-3, f1^-2*f2^-3*f1^-1, f1^-1*f2^-3*f1^-2, f2^-1*f1^-3*f2^-2,
f2^-2*f1^-3*f2^-1, f2^-3*f1^-3, f1^-3*f2^3, f2^-3*f1^3,
f1^-2*f2^3*f1^-1, f2^-2*f1^3*f2^-1, f1^-1*f2^3*f1^-2, f2^-1*f1^3*f2^-2,
f1*f2^-3*f1^2, f2*f1^-3*f2^2, f1^2*f2^-3*f1, f2^2*f1^-3*f2, f1^3*f2^-3,
f2^3*f1^-3, f1^3*f2^3, f1^2*f2^3*f1, f1*f2^3*f1^2, f2*f1^3*f2^2,
f2^2*f1^3*f2, f2^3*f1^3],
[f1^-2*f2^-1*f1^-2*f2, f1^-1*f2^-1*f1^-2*f2*f1^-1,
f2^-1*f1^-1*f2^-2*f1*f2^-1, f2^-2*f1^-1*f2^-2*f1, f1^-1*f2^-2*f1*f2^-2,
f2^-1*f1^-2*f2*f1^-2, f1^-2*f2*f1^-2*f2^-1, f2^-2*f1*f2^-2*f1^-1,
f1^-1*f2^2*f1^-2*f1^-1, f2^-1*f1*f2^-2*f1^-1*f2^-1,
f1^-1*f2^2*f1*f2^2, f2^-1*f1^2*f2*f1^2, f1*f2^2*f1^-1*f2^-2,
f2*f1^-2*f2^-1*f1^-2, f1*f2^-1*f1^2*f2*f1, f2*f1^-1*f2^2*f1*f2,
f1^2*f2^-1*f1^2*f2, f2^2*f1^-1*f2^2*f1, f1*f2^2*f1^-1*f2^2,
f2*f1^2*f2^-1*f1^2, f1^2*f2*f1^2*f2^-1, f1*f2*f1^2*f2^-1*f1,
f2*f1*f2^2*f1^-1*f2, f2^2*f1*f2^2*f1^-1],
[f1^-2*f2^-1*f1^-1*f2^2, f1^-2*f2^-2*f1*f2^-1, f1^-1*f2^-2*f1^-2*f2,
f2^-1*f1^-2*f2^-2*f1, f2^-2*f1^-2*f2*f1^-1, f2^-2*f1^-1*f2^-1*f1^2,
f1^-1*f2^-1*f1^-1*f2^2*f1^-1, f1^-1*f2^-2*f1*f2^-1*f1^-1,
f2^-1*f1^-2*f2*f1^-1*f2^-1, f2^-1*f1^-1*f2^-1*f1^2*f2^-1,
f1^-2*f2*f1^-1*f2^-2, f2^-2*f1*f2^-1*f1^-2, f1^-2*f2^2*f1*f2,
f1^-1*f2^-1*f1^2*f2^-2, f2^-1*f1^-1*f1^2*f2^2*f1^-2, f2^-2*f1^2*f2*f1,
f1^-1*f2^-1*f1^2*f1^-1, f2^-1*f1*f2^2*f1^-1*f2^-1,
f1^-1*f2^2*f1^-2*f2^-1, f2^-1*f1^2*f2^-2*f1^-1, f1^-1*f2*f1^2*f2^2,
f1^-1*f2^2*f1*f2*f1^-1, f2^-1*f1^2*f2*f1*f2^-1, f2^-1*f1*f2^2*f1^2,
f1*f2^-1*f1^-2*f2^-2, f1*f2^-2*f1^-1*f2^-1*f1, f2*f1^-2*f2^-1*f1^-1*f2,
f2*f1^-1*f2^-2*f1^-2, f1*f2^-2*f1^2*f2, f2*f1^-2*f2^2*f1,
f1*f2^-1*f1*f2^2*f1, f2*f1^-1*f2*f1^2*f2, f1^2*f2^-2*f1^-1*f2^-1,
f1*f2*f1^-2*f2^2, f2*f1*f2^-2*f1^2, f2^2*f1^-2*f2^-1*f1^-1,
f1^2*f2^-1*f1*f2^2, f2^2*f1^-1*f2*f1^2, f1*f2*f1*f2^-2*f1,
f1*f2^2*f1^-1*f2*f1, f2*f1^2*f2^-1*f1*f2, f2*f1*f2*f1^-2*f2,
f1^2*f2*f1*f2^-2, f1^2*f2^2*f1^-1*f2, f1*f2^2*f1^2*f2^-1,
f2*f1^2*f2^2*f1^-1, f2^2*f1^2*f2^-1*f1, f2^2*f1*f2*f1^-2],
[f1^-2*f2^-2*f1^-1*f2^2, f1^-1*f2^-1*f1^-1*f2^-2*f2^2, f1^-1*f2^-2*f1^2*f1^-1*f2*f1^-1,
f2^-1*f1^-2*f2^-1*f1*f2^-1, f2^-1*f1^-1*f2^-1*f2^-1*f1^2,
f2^-2*f1^-2*f2^-1*f1, f1^-2*f2^-1*f1*f2^-2, f1^-1*f2^-2*f1^2*f2^-1,
f2^-1*f1^-2*f2^2*f1^-1, f2^-2*f1^-1*f2*f1^-2,
f1^-1*f2^-1*f1*f1*f2^-2*f1^-1, f2^-1*f1^-1*f2*f1^-2*f2^-1,
f1^-2*f2*f1*f2^2, f1^-2*f2^2*f1^-1*f2^-1, f2^-2*f1^2*f2^-1*f1^-1,
f2^-2*f1*f2*f1^2, f1^-1*f2*f1^-2*f2^-2, f2^-1*f1*f2^-2*f1^-2,
f1^-1*f2^2*f1^-1*f2^-1*f1^-1, f2^-1*f1^2*f2^-1*f1^-1*f2^-1,
f1^-1*f2*f1*f2^2*f1^-1, f1^-1*f2^2*f1^2*f2, f2^-1*f1^2*f2^2*f1,
f2^-1*f1*f2*f1^2*f2^-1, f1*f2^-1*f1^-1*f2^-2*f1, f1*f2^-2*f1^-2*f2^-1,
f2*f1^-2*f2^-2*f1^-1, f2*f1^-1*f2^-1*f1^2, f1*f2^-2*f1*f2*f1,
f1*f2^2*f1^2*f2^-1, f2^2*f1^2*f2^-1*f1, f1*f2*f1*f2^-2*f1],

```
f2*f1^-2*f2*f1*f2 ,  f1*f2^-1*f1^2*f2^2,  f2*f1^-1*f2^2*f1^2,
f1^2*f2^-1*f1^-1*f2^-2,  f1^2*f2^-2*f1*f2 ,  f2^2*f1^-2*f2*f1 ,
f2^2*f1^-1*f2^-1*f1^-2,  f1*f2*f1^-1*f2^2*f1 ,  f2*f1*f2^-1*f1^2*f2 ,
f1^2*f2*f1^-1*f2^2,  f1*f2^2*f1^-2*f2 ,  f2*f1^2*f2^-2*f1 ,
f2^2*f1*f2^-1*f1^2,  f1^2*f2^2*f1*f2^-1,  f1*f2*f1^2*f2^-2,
f1*f2^2*f1*f2^-1*f1 ,  f2*f1^2*f2*f1^-1*f2 ,  f2*f1*f2^2*f1^-2,
f2^2*f1^2*f2*f1^-1 ],
[ f1^-2*f2^-2*f1*f2 ,  f2^-2*f1^-2*f2*f1 ,  f1^-2*f2^-1*f1*f2^2,
f1^-1*f2^-2*f1^2*f2 ,  f1^-1*f2^-2*f1*f2*f1^-1,  f2^-1*f1^-2*f2*f1*f2^-1,
f2^-1*f1^-2*f2^2*f1 ,  f2^-2*f1^-1*f2*f1^2,  f1^-2*f2*f1*f2^-2,
f1^-2*f2^2*f1*f2^-1,  f1^-1*f2^-1*f1^2*f2^2,  f1^-1*f2^-1*f1*f2^2*f1^-1,
f2^-1*f1^-1*f2*f1^2*f2^-1,  f2^-1*f1^-1*f2^2*f1^2,  f2^-2*f1^2*f2*f1^-1,
f2^-2*f1*f2*f1^-2,  f1^-1*f2*f1*f2^-2*f1^-1,  f2^-1*f1*f2*f1^-2*f2^-1,
f1^-1*f2*f1^2*f2^-2,  f1^-1*f2^2*f1^2*f2^-1,  f1^-1*f2^2*f1*f2^-1*f1^-1,
f2^-1*f1^2*f2*f1^-1*f2^-1,  f2^-1*f1^2*f2^2*f1^-1,  f2^-1*f1*f2^2*f1^-2,
f1*f2^-1*f1^-2*f2^2,  f1*f2^-2*f1^-2*f2 ,  f1*f2^-2*f1^-1*f2*f1 ,
f2*f1^-2*f2^-1*f1*f2 ,  f2*f1^-1*f2^-2*f1 ,  f2*f1^-1*f2^-2*f1^2,
f1*f2^-1*f1^-1*f2^2*f1 ,  f2*f1^-1*f2^-1*f1^2*f2 ,  f1^2*f2^-1*f1^-1*f2^2,
f1^2*f2^-2*f1^-1*f2 ,  f1*f2*f1^-2*f2^-2 ,  f1*f2*f1^-1*f2^-2*f1 ,
f2*f1*f2^-1*f1^-2*f2 ,  f2*f1*f2^-2*f1^-2,  f2^2*f1^-2*f2^-1*f1 ,
f2^2*f1^-1*f2^-1*f1^2,  f1^2*f2*f1^-1*f2^-2,  f1*f2^2*f1^-2*f2^-1,
f1*f2^2*f1^-1*f2^-1*f1 ,  f2*f1^2*f2^-1*f1^-1*f2 ,  f2*f1^2*f2^-2*f1^-1,
f2^2*f1*f2^-1*f1^-2,  f1^2*f2^2*f1^-1*f2^-1,  f2^2*f1^2*f2^-1*f1^-1 ],
[ f1^-2*f2^-1*f1^2*f2 ,  f1^-1*f2^-2*f1*f2^2,  f2^-1*f1^-2*f2*f1^2,
f2^-2*f1^-1*f2^2*f1 ,  f1^-2*f2*f1^2*f2^-1,  f1^-1*f2^-1*f1^2*f2*f1^-1,
f2^-1*f1^-1*f2^2*f1*f2^-1,  f2^-2*f1*f2^2*f1^-1,
f1^-1*f2*f1^2*f2^-1*f1^-1,  f1^-1*f2^2*f1*f2^-2,  f2^-1*f1^2*f2*f1^-2,
f2^-1*f1^2*f2^2*f1^-1*f1^-1,  f1^-1*f2^2*f1*f2^-2,  f2^-1*f1^2*f2*f1^-1*f2^-2,
f2*f1^-2*f2^-1*f1^2,  f2*f1^-1*f2^-2*f1*f2 ,  f1^2*f2^-1*f1^-2*f2 ,
f1*f2*f1^-2*f2^-1*f1 ,  f2*f1*f2^-2*f1^-1*f2 ,  f2^2*f1^-1*f2^-2*f1 ,
f1^2*f2*f1^-2*f2^-1,  f1*f2^2*f1^-1*f2^-2,  f2*f1^2*f2^-1*f1^-2,
f2^2*f1*f2^-2*f1^-1 ] ].
```

B.2   Equivalence classes of minimal words of lengths $2, 3, 4$ and 5 in F_3

EquivCl(3,2):=[[f1^-2, f2^-2, f3^-2, f1^2, f2^2, f3^2]].

EquivCl(3,3):=[[f1^-3, f2^-3, f3^-3, f1^3, f2^3, f3^3]].

EquivCl(3,4):=[[f1^-4, f2^-4, f3^-4, f1^4, f2^4, f3^4],
[f1^-2*f2^-2, f1^-2*f3^-2, f1^-1*f2^-2*f1^-1, f1^-1*f3^-2*f1^-1,
f2^-1*f1^-2*f2^-1, f2^-2*f1^-2, f2^-2*f3^-2, f2^-1*f3^-2*f2^-1,
f3^-1*f1^-2*f3^-1, f3^-1*f2^-2*f3^-1, f3^-2*f1^-2, f3^-2*f2^-2,
f1^-1*f2^-1*f1^-1*f2, f1^-1*f3^-1*f1^-1*f3, f2^-1*f1^-1*f1^-1*f2^-1*f1,
f2^-1*f3^-1*f2^-1*f3, f3^-1*f1^-1*f3^-1*f1, f3^-1*f2^-1*f3^-1*f2,
f1^-1*f2^-1*f1*f2^-1, f1^-1*f3^-1*f1*f3^-1, f2^-1*f1^-1*f2*f1^-1,
f2^-1*f3^-1*f2*f3^-1, f3^-1*f1^-1*f3*f1^-1, f3^-1*f2^-1*f3*f2^-1,
```
```

f1^−2∗f2^2, f1^−2∗f3^2, f2^−2∗f1^2, f2^−2∗f3^2, f3^−2∗f1^2,
f3^−2∗f2^2, f1^−1∗f2∗f1^−1∗f2^−1, f1^−1∗f3∗f1^−1∗f3^−1,
f2^−1∗f1∗f2^−1∗f1^−1, f2^−1∗f3∗f2^−1∗f3^−1, f3^−1∗f1∗f3^−1∗f1^−1,
f3^−1∗f2∗f3^−1∗f2^−1, f1^−1∗f2^2∗f1^−1, f1^−1∗f3^2∗f1^−1,
f2^−1∗f1^2∗f2^−1, f2^−1∗f3^2∗f2^−1, f3^−1∗f1^2∗f3^−1, f3^−1∗f2^2∗f3^−1,
f1^−1∗f2∗f1∗f2, f1^−1∗f3∗f1∗f3, f2^−1∗f1∗f2∗f1, f2^−1∗f3∗f2∗f3,
f3^−1∗f1∗f3∗f1, f3^−1∗f2∗f3∗f2, f1∗f2^−1∗f1^−1∗f2^−1,
f1∗f3^−1∗f1^−1∗f3^−1, f2∗f1^−1∗f2^−1∗f1^−1, f2∗f3^−1∗f2^−1∗f3^−1,
f3∗f1^−1∗f3^−1∗f1^−1, f3∗f2^−1∗f3^−1∗f2^−1, f1∗f2^−2∗f1, f1∗f3^−2∗f1,
f2∗f1^−2∗f2, f2∗f3^−2∗f2, f3∗f1^−2∗f3, f3∗f2^−2∗f3, f1∗f2^−1∗f1∗f2,
f1∗f3^−1∗f1∗f3, f2∗f1^−1∗f2∗f1, f2∗f3^−1∗f2∗f3, f3∗f1^−1∗f3∗f1,
f3∗f2^−1∗f3∗f2, f1^2∗f2^−2, f1^2∗f3^−2, f2^2∗f1^−2, f2^2∗f3^−2,
f3^2∗f1^−2, f3^2∗f2^−2, f1∗f2∗f1^−1∗f2, f1∗f3∗f1^−1∗f3, f2∗f1∗f2^−1∗f1,
f2∗f3∗f2^−1∗f3, f3∗f1∗f3^−1∗f1, f3∗f2∗f3^−1∗f2, f1∗f2∗f1∗f2^−1,
f1∗f3∗f1∗f3^−1, f2∗f1∗f2∗f1^−1, f2∗f3∗f2∗f3^−1, f3∗f1∗f3∗f1^−1,
f3∗f2∗f3∗f2^−1, f1^2∗f2^2, f1^2∗f3^2, f1∗f2^2∗f1, f1∗f3^2∗f1,
f2∗f1^2∗f2, f2^2∗f1^2, f2^2∗f3^2, f2∗f3^2∗f2, f3∗f1^2∗f3, f3∗f2^2∗f3,
f3^2∗f1^2, f3^2∗f2^2],
[f1^−1∗f2^−1∗f1∗f2, f1^−1∗f3^−1∗f1∗f3,
f2^−1∗f1^−1∗f2∗f1, f2^−1∗f3^−1∗f2∗f3, f3^−1∗f1^−1∗f3∗f1,
f3^−1∗f2^−1∗f3∗f2, f1^−1∗f2∗f1∗f2^−1, f1^−1∗f3∗f1∗f3^−1,
f2^−1∗f1∗f2∗f1^−1, f2^−1∗f3∗f2∗f3^−1, f3^−1∗f1∗f3∗f1^−1,
f3^−1∗f2∗f3∗f2^−1, f1∗f2^−1∗f1^−1∗f2, f1∗f3^−1∗f1^−1∗f3,
f2∗f1^−1∗f2^−1∗f1, f2∗f3^−1∗f2^−1∗f3, f3∗f1^−1∗f3^−1∗f1,
f3∗f2^−1∗f3^−1∗f2, f1∗f2∗f1^−1∗f2^−1, f1∗f3∗f1^−1∗f3^−1,
f2∗f1∗f2^−1∗f1^−1, f2∗f3∗f2^−1∗f3^−1, f3∗f1∗f3^−1∗f1^−1,
f3∗f2∗f3^−1∗f2^−1]].

EquivCl(3,5):=[[f1^−5, f2^−5, f3^−5, f1^5, f2^5, f3^5],
[f1^−3∗f2^−2, f1^−3∗f3^−2, f1^−2∗f2^−2∗f1^−1, f1^−2∗f2^−3,
f1^−2∗f3^−2∗f1^−1, f1^−2∗f3^−3, f1^−1∗f2^−2∗f1^−2, f1^−1∗f2^−2∗f1^−1∗f2,
f1^−1∗f2^−3∗f1^−1, f1^−1∗f3^−2∗f1^−2, f1^−1∗f3^−2∗f1^−1∗f3,
f1^−1∗f3^−3∗f1^−1, f2^−1∗f1^−3∗f2^−1, f2^−1∗f1^−2∗f2^−1∗f1,
f2^−1∗f1^−2∗f2^−2, f2^−2∗f1^−3, f2^−2∗f1^−2∗f2^−1, f2^−3∗f1^−2,
f2^−3∗f3^−2, f2^−2∗f3^−2∗f2^−1, f2^−2∗f3^−3, f2^−1∗f3^−2∗f2^−2,
f2^−1∗f3^−2∗f2^−1∗f3, f2^−1∗f3^−3∗f2^−1, f3^−1∗f1^−3∗f3^−1,
f3^−1∗f1^−2∗f3^−1∗f1, f3^−1∗f1^−2∗f3^−2, f3^−1∗f2^−3∗f3^−1,
f3^−1∗f2^−2∗f3^−1∗f2, f3^−1∗f2^−2∗f3^−2, f3^−2∗f1^−3, f3^−2∗f1^−2∗f3^−1,
f3^−2∗f2^−3, f3^−2∗f2^−2∗f3^−1, f3^−3∗f1^−2, f1^−3∗f2^2,
f1^−3∗f3^2, f1^−2∗f2^−1∗f1∗f2^−1, f1^−2∗f3^−1∗f1∗f3^−1,
f1^−1∗f2^−1∗f1^−1∗f2^2, f1^−1∗f3^−1∗f1^−1∗f3^2, f2^−1∗f1^−1∗f2^−1∗f1^2,
f2^−2∗f1^−1∗f2∗f1^−1, f2^−3∗f1^2, f2^−3∗f3^2, f2^−2∗f3^−1∗f2∗f3^−1,
f2^−1∗f3^−1∗f2^−1∗f3^2, f3^−1∗f1^−1∗f3^−1∗f1^2, f3^−1∗f2^−1∗f3^−1∗f2^2,
f3^−2∗f1^−1∗f3∗f1^−1, f3^−2∗f2^−1∗f3∗f2^−1, f3^−3∗f1^2, f3^−3∗f2^2,
f1^−1∗f2^−1∗f1∗f2^−1∗f1^−1, f1^−1∗f3^−1∗f1∗f3^−1∗f1^−1,
f2^−1∗f1^−1∗f2∗f1^−1∗f2^−1, f2^−1∗f3^−1∗f2∗f3^−1∗f2^−1,
f3^−1∗f1^−1∗f3∗f1^−1∗f3^−1, f3^−1∗f2^−1∗f3∗f2^−1∗f3^−1, f1^−2∗f2∗f1∗f2 ,
f1^−2∗f2^2∗f1^−1, f1^−2∗f2^3, f1^−2∗f3∗f1∗f3, f1^−2∗f3^2∗f1^−1,

f1^-2*f3^3, f1^-1*f2^-1*f1^2*f2^-1, f1^-1*f3^-1*f1^2*f3^-1,
f2^-1*f1^-1*f2^2*f1^-1, f2^-2*f1^3, f2^-2*f1^2*f2^-1, f2^-2*f1*f2*f1,
f2^-2*f3*f2*f3, f2^-2*f3^2*f2^-1, f2^-2*f3^3, f2^-1*f3^-1*f2^2*f3^-1,
f3^-1*f1^-1*f3^2*f1^-1, f3^-1*f2^-1*f3^2*f2^-1, f3^-2*f1^3,
f3^-2*f1^2*f3^-1, f3^-2*f1*f3*f1, f3^-2*f2^3, f3^-2*f2^2*f3^-1,
f3^-2*f2*f3*f2, f1^-1*f2*f1^-1*f2^-2, f1^-1*f3*f1^-1*f3^-2,
f2^-1*f1*f2^-1*f1^-2, f2^-1*f3*f2^-1*f3^-2, f3^-1*f1*f3^-1*f1^-2,
f3^-1*f2*f3^-1*f2^-2, f1^-1*f2^2*f1^-2, f1^-1*f2^2*f1^-1*f2^-1,
f1^-1*f3^2*f1^-2, f1^-1*f3^2*f1^-1*f3^-1, f2^-1*f1^2*f2^-1*f1^-1,
f2^-1*f1^2*f2^-2, f2^-1*f3^2*f2^-2, f2^-1*f3^2*f2^-1*f3^-1,
f3^-1*f1^2*f3^-1*f1^-1, f3^-1*f1^2*f3^-2, f3^-1*f2^2*f3^-1*f2^-1,
f3^-1*f2^2*f3^-2, f1^-1*f2*f1^2*f2, f1^-1*f2*f1*f2*f1^-1,
f1^-1*f2^3*f1^-1, f1^-1*f3*f1^2*f3, f1^-1*f3*f1*f3*f1^-1,
f1^-1*f3^3*f1^-1, f2^-1*f1^3*f2^-1, f2^-1*f1*f2*f1*f2^-1,
f2^-1*f1*f2^2*f1, f2^-1*f3*f2^2*f3, f2^-1*f3*f2*f3*f2^-1,
f2^-1*f3^3*f2^-1, f3^-1*f1^3*f3^-1, f3^-1*f1*f3*f1*f3^-1,
f3^-1*f1*f3^2*f1, f3^-1*f2^3*f3^-1, f3^-1*f2*f3*f2*f3^-1,
f3^-1*f2*f3^2*f2, f1*f2^-1*f1^-2*f2^-1, f1*f2^-1*f1^-1*f2^-1*f1,
f1*f2^-3*f1, f1*f3^-1*f1^-2*f3^-1, f1*f3^-1*f1^-1*f3^-1*f1,
f1*f3^-3*f1, f2*f1^-3*f2, f2*f1^-1*f2^-1*f1^-1*f2, f2*f1^-1*f2^-2*f1^-1,
f2*f3^-1*f2^-2*f3^-1, f2*f3^-1*f2^-1*f3^-1*f2, f2*f3^-3*f2,
f3*f1^-3*f3, f3*f1^-1*f3^-1*f1^-1*f3, f3*f1^-1*f3^-2*f1^-1,
f3*f2^-3*f3, f3*f2^-1*f3^-1*f2^-1*f3, f3*f2^-1*f3^-2*f2^-1,
f1*f2^-2*f1^2, f1*f2^-2*f1*f2, f1*f3^-2*f1^2, f1*f3^-2*f1*f3,
f2*f1^-2*f2*f1, f2*f1^-2*f2^2, f2*f3^-2*f2^2, f2*f3^-2*f2*f3,
f3*f1^-2*f3*f1, f3*f1^-2*f3^2, f3*f2^-2*f3*f2, f3*f2^-2*f3^2,
f1*f2^-1*f1*f2^2, f1*f3^-1*f1*f3^2, f2*f1^-1*f2*f1^2, f2*f3^-1*f2*f3^2,
f3*f1^-1*f3*f1^2, f3*f2^-1*f3*f2^2, f1^2*f2^-1*f1^-1*f2^-1,
f1^2*f2^2*f2^-2*f1, f1^2*f3^-1*f1^-1*f3^-1, f1^2*f3^-2*f1,
f1^2*f3^-3, f1*f3*f1^-2*f2, f1*f3*f1^-2*f3, f2*f1*f2^-2*f1, f2^2*f1^-3,
f2^2*f1^-2*f2, f2^2*f1^-1*f2^-1*f1^-1, f2^2*f3^-1*f2^-1*f3^-1,
f2^2*f3^-2*f2, f2^2*f3^-3, f2*f3*f2^-2*f3, f3*f1*f3^-2*f1,
f3*f2*f3^-2*f2, f3^2*f1^-3, f3^2*f1^-2*f3, f3^2*f1^-1*f3^-1*f1^-1,
f3^2*f2^-3, f3^2*f2^-2*f3, f3^2*f2^-1*f3^-1*f2^-1, f1*f2*f1^-1*f2*f1,
f1*f3*f1^-1*f3*f1, f2*f1*f2^-1*f1*f2, f2*f3*f2^-1*f3*f2,
f3*f1*f3^-1*f1*f3, f3*f2*f3^-1*f2*f3, f1^3*f2^-2, f1^3*f3^-2,
f1^2*f2*f1^-1*f2, f1^2*f3*f1^-1*f3, f1*f2*f1*f2^-2, f1*f3*f1*f3^-2,
f2*f1*f2*f1^-2, f2^2*f1*f2^-1*f1, f2^3*f1^-2, f2^3*f3^-2,
f2^2*f3*f2^-1*f3, f2*f3*f2*f3^-2, f3*f1*f3*f1^-2, f3*f2*f3*f2^-2,
f3^2*f1*f3^-1*f1, f3^2*f2*f3^-1*f2, f3^3*f1^-2, f1^3*f2^2,
f1^3*f3^2, f1^2*f2*f2^2*f1, f1^2*f2^3, f1^2*f3^2*f1, f1^2*f3^3,
f1*f2^2*f1^2, f1*f2^2*f1*f2^-1, f1*f2^3*f1, f1*f3^2*f1^2,
f1*f3^2*f1*f3^-1, f1*f3^3*f1, f2*f1^3*f2, f2*f1^2*f2*f1^-1,
f2*f1^2*f2^2, f2^2*f1^3, f2^2*f1^2*f2, f2^3*f1^2, f2^3*f3^2,
f2^2*f3^2*f2, f2^2*f3^3, f2*f3^2*f2^2, f2*f3^2*f2*f3^-1, f2*f3^3*f2,
f3*f1^3*f3, f3*f1^2*f3*f1^-1, f3*f1^2*f3^2, f3*f2^3*f3,
f3*f2^2*f3*f2^-1, f3*f2^2*f3^2, f3^2*f1^3, f3^2*f1^2*f3, f3^2*f2^3,
f3^2*f2^2*f3, f3^3*f1^2, f3^3*f2^2],

```
[  f1^-2*f2^-1*f1^-1*f2 ,  f1^-2*f3^-1*f1^-1*f3 ,  f1^-1*f2^-1*f1^-2*f2 ,
f1^-1*f3^-1*f1^-2*f3 ,  f2^-1*f1^-1*f2^-2*f1 ,  f2^-2*f1^-1*f2^-1*f1 ,
f2^-2*f3^-1*f2^-1*f3 ,  f2^-1*f3^-1*f2^-2*f3 ,  f3^-1*f1^-1*f3^-2*f1 ,
f3^-1*f2^-1*f3^-2*f2 ,  f3^-2*f1^-1*f3^-1*f1 ,  f3^-2*f2^-1*f3^-1*f2 ,
f1^-1*f2^-1*f1^-1*f2*f1^-1 ,  f1^-1*f2^-2*f1*f2^-1 ,
f1^-1*f3^-1*f1^-1*f3*f1^-1 ,  f1^-1*f3^-2*f1*f3^-1 ,  f2^-1*f1^-1*f2*f1^-1 ,
f2^-1*f1^-1*f2^-1*f1*f2^-1 ,  f2^-1*f3^-1*f2^-1*f3*f2^-1 ,
f2^-1*f3^-2*f2*f3^-1 ,  f3^-1*f1^-1*f2*f3*f1^-1 ,  f3^-1*f1^-1*f3^-1*f1*f3^-1 ,
f3^-1*f2^-2*f3*f2^-1 ,  f3^-1*f2^-1*f3^-1*f2*f3^-1 ,  f1^-2*f2*f1^-1*f2^-1 ,
f1^-2*f3*f1^-1*f3^-1 ,  f1^-1*f2^-1*f1*f2^-2 ,  f1^-1*f3^-1*f1*f3^-2 ,
f2^-1*f1^-1*f2*f1^-2 ,  f2^-2*f1*f2^-1*f1^-1 ,  f2^-2*f3*f2^-1*f3^-1 ,
f2^-1*f3^-1*f2*f3^-2 ,  f3^-1*f1^-1*f3*f1^-2 ,  f3^-1*f2^-1*f3*f2^-2 ,
f3^-2*f1*f3^-1*f1^-1 ,  f3^-2*f2*f3^-1*f2^-1 ,  f1^-1*f2*f1^-2*f2^-1 ,
f1^-1*f2*f1^-1*f2^-1*f1^-1 ,  f1^-1*f3*f1^-2*f3^-1 ,
f1^-1*f3*f1^-1*f3^-1*f1^-1 ,  f2^-1*f1*f2^-1*f1^-1*f2^-1 ,
f2^-1*f1*f2^-2*f1^-1 ,  f2^-1*f3*f2^-2*f3^-1 ,  f2^-1*f3*f2^-1*f3^-1*f2^-1 ,
f3^-1*f1*f3^-1*f1^-1*f3^-1 ,  f3^-1*f1*f3^-2*f1^-1 ,
f3^-1*f2*f3^-1*f2^-1*f3^-1 ,  f3^-1*f2*f3^-2*f2^-1 ,  f1^-1*f2*f1*f2^2 ,
f1^-1*f2^2*f1*f2 ,  f1^-1*f3*f1*f3^2 ,  f1^-1*f3^2*f1*f3 ,  f2^-1*f1^2*f2*f1 ,
f2^-1*f1*f2*f1^2 ,  f2^-1*f3*f2*f3^2 ,  f2^-1*f3^2*f2*f3 ,  f3^-1*f1^2*f3*f1 ,
f3^-1*f1*f3*f1^2 ,  f3^-1*f2^2*f3*f2 ,  f3^-1*f2*f3*f2^2 ,
f1*f2^-1*f1^-1*f2^-2 ,  f1*f2^-2*f1^-1*f2^-1 ,  f1*f3^-1*f1^-1*f3^-2 ,
f1*f3^-2*f1^-1*f3^-1 ,  f2*f1^-2*f2^-1*f1^-1 ,  f2*f1^-1*f2^-1*f1^-2 ,
f2*f3^-1*f2^-1*f3^-2 ,  f2*f3^-2*f2^-1*f3^-1 ,  f3*f1^-2*f3^-1*f1^-1 ,
f3*f1^-1*f3^-1*f1^-2 ,  f3*f2^-2*f3^-1*f2^-1 ,  f3*f2^-1*f3^-1*f2^-2 ,
f1*f2^-1*f1^2*f2 ,  f1*f2^-1*f1*f2*f1 ,  f1*f3^-1*f1^2*f3 ,
f1*f3^-1*f1*f3*f1 ,  f2*f1^-1*f2*f1*f2 ,  f2*f1^-1*f2^2*f1 ,
f2*f3^-1*f2^2*f3 ,  f2*f3^-1*f2*f3*f2 ,  f3*f1^-1*f3*f1*f3 ,
f3*f1^-1*f3^2*f1 ,  f3*f2^-1*f3*f2*f3 ,  f3*f2^-1*f3^2*f2 ,  f1^2*f2^-1*f1*f2 ,
f1^2*f3^-1*f1*f3 ,  f1*f2*f1^-1*f2^2 ,  f1*f3*f1^-1*f3^2 ,  f2*f1*f2^-1*f1^2 ,
f2^2*f1^-1*f2*f1 ,  f2^2*f3^-1*f2*f3 ,  f2*f3*f2^-1*f3^2 ,  f3*f1*f3^-1*f1^2 ,
f3*f2*f3^-1*f2^2 ,  f1*f2*f1*f2^-1*f1 ,
f1*f2^2*f1^-1*f2 ,  f1*f3*f1*f3^-1*f1 ,  f1*f3^2*f1^-1*f3 ,  f2*f1^2*f2^-1*f1 ,
f2*f1*f2*f1^-1*f2 ,  f2*f3*f2*f3^-1*f2 ,  f2*f3^2*f2^-1*f3 ,
f3*f1^2*f3^-1*f1 ,  f3*f1*f3*f1^-1*f3 ,  f3*f2^2*f3^-1*f2 ,
f3*f2*f3*f2^-1*f3 ,  f1^2*f2*f1*f2^-1 ,  f1^2*f3*f1*f3^-1 ,  f1*f2*f1^2*f2^-1 ,
f1*f3*f1^2*f3^-1 ,  f2*f1*f2^2*f1^-1 ,  f2^2*f1*f2*f1^-1 ,  f2^2*f3*f2*f3^-1 ,
f2*f3*f2^2*f3^-1 ,  f3*f1*f3^2*f1^-1 ,  f3*f2*f3^2*f2^-1 ,  f3^2*f1*f3*f1^-1 ,
f3^2*f2*f3*f2^-1 ],
[  f1^-2*f2^-1*f1*f2 ,  f1^-2*f3^-1*f1*f3 ,  f1^-1*f2^-2*f1*f2 ,
f1^-1*f3^-2*f1*f3 ,  f2^-1*f1^-2*f2*f1 ,  f2^-2*f1^-1*f2*f1 ,
f2^-2*f3^-1*f2*f3 ,  f2^-1*f3^-2*f2*f3 ,  f3^-1*f1^-2*f3*f1 ,
f3^-1*f2^-2*f3*f2 ,  f3^-2*f1^-1*f3*f1 ,  f3^-2*f2^-1*f3*f2 ,
f1^-2*f2*f1*f2^-1 ,  f1^-2*f3*f1*f3^-1 ,  f1^-1*f2^-1*f1^2*f2 ,
f1^-1*f2^-1*f1*f2*f1^-1 ,  f1^-1*f2^-1*f1*f2^2 ,  f1^-1*f3^-1*f1^2*f3 ,
f1^-1*f3^-1*f1*f3*f1^-1 ,  f1^-1*f3^-1*f1*f3^2 ,  f2^-1*f1^-1*f2*f1^2 ,
f2^-1*f1^-1*f2*f1*f2^-1 ,  f2^-1*f1^-1*f2^2*f1 ,  f2^-2*f1*f2*f1^-1 ,
f2^-2*f3*f2*f3^-1 ,  f2^-1*f3^-1*f2^2*f3 ,  f2^-1*f3^-1*f2*f3*f2^-1 ,
```

f2^-1*f3^-1*f2*f3^2, f3^-1*f1^-1*f3*f1^2, f3^-1*f1^-1*f3*f1*f3^-1,
f3^-1*f1^-1*f3^2*f1, f3^-1*f2^-1*f3*f2^2, f3^-1*f2^-1*f3*f2*f3^-1,
f3^-1*f2^-1*f3^2*f2, f3^-2*f1*f3*f1^-1, f3^-2*f2*f3*f2^-1,
f1^-1*f2*f1*f2^-1*f1^-1, f1^-1*f2*f1*f2^-2, f1^-1*f3*f1*f3^-1*f1^-1,
f1^-1*f3*f1*f3^-2, f2^-1*f1*f2*f1^-2, f2^-1*f1*f2*f1^-1*f2^-1,
f2^-1*f3*f2*f3^-1*f2^-1, f2^-1*f3*f2*f3^-2, f3^-1*f1*f3*f1^-2,
f3^-1*f1*f3*f1^-1*f3^-1, f3^-1*f2*f3*f2^-2, f3^-1*f2*f3*f2^-1*f3^-1,
f1^-1*f2*f1^2*f2^-1, f1^-1*f2^2*f1*f2^-1, f1^-1*f3*f1^2*f3^-1,
f1^-1*f3^2*f1*f3^-1, f2^-1*f1^2*f2*f1^-1, f2^-1*f1*f2^2*f1^-1,
f2^-1*f3*f2^2*f3^-1, f2^-1*f3^2*f2*f3^-1, f3^-1*f1^2*f3*f1^-1,
f3^-1*f1*f3^2*f1^-1, f3^-1*f2^2*f3*f2^-1, f3^-1*f2*f3^2*f2^-1,
f1*f2^-1*f1^-2*f2, f1*f2^-2*f1^-1*f2, f1*f3^-1*f1^-2*f3 ,
f1*f3^-2*f1^-1*f3, f2*f1^-2*f2^-1*f1, f2*f1^-1*f2^-2*f1 ,
f2*f3^-1*f2^-2*f3, f2*f3^-2*f2^-1*f3, f3*f1^-2*f3^-1*f1 ,
f3*f1^-1*f3^-2*f1, f3*f2^-2*f3^-1*f2, f3*f2^-1*f3^-2*f2 ,
f1*f2^-1*f1^-1*f2*f1, f1*f2^-1*f1^-1*f2^2, f1*f3^-1*f1^-1*f3*f1,
f1*f3^-1*f1^-1*f3^2, f2*f1^-1*f2^-1*f1^2, f2*f1^-1*f2^-1*f1*f2,
f2*f3^-1*f2^-1*f3*f2, f2*f3^-1*f2^-1*f3^2, f3*f1^-1*f3^-1*f1^2,
f3*f1^-1*f3^-1*f1*f3, f3*f2^-1*f3^-1*f2^2, f3*f2^-1*f3^-1*f2*f3 ,
f1^2*f2^-1*f1^-1*f2, f1^2*f3^-1*f1^-1*f3, f1*f2*f1^-2*f2^-1,
f1*f2*f1^-1*f2^-1*f1, f1*f2*f1^-1*f2^-2, f1*f3*f1^-2*f3^-1,
f1*f3*f1^-1*f3^-1*f1, f1*f3*f1^-1*f3^-2, f2*f1*f2^-1*f1^-2,
f2*f1*f2^-1*f1^-1*f2, f2*f1*f2^-2*f1^-1, f2^2*f1^-1*f2^-1*f1 ,
f2^2*f3^-1*f2^-1*f3, f2*f3*f2^-2*f3^-1, f2*f3*f2^-1*f3^-1*f2,
f2*f3*f2^-1*f3^-2, f3*f1*f3^-1*f1^-2, f3*f1*f3^-1*f1^-1*f3 ,
f3*f1*f3^-2*f1^-1, f3*f2*f3^-1*f2^-2, f3*f2*f3^-1*f2^-1*f3 ,
f3*f2*f3^-2*f2^-1, f3^2*f1^-1*f3^-1*f1, f3^2*f2^-1*f3^-1*f2 ,
f1^2*f2*f1^-1*f2^-1, f1^2*f3*f1^-1*f3^-1, f1*f2^2*f1^-1*f2^-1,
f1*f3^2*f1^-1*f3^-1, f2*f1^2*f2^-1*f1^-1, f2^2*f3*f2^-1*f3^-1,
f2^2*f3*f2^-1*f3^-1, f2*f3^2*f2^-1*f3^-1, f3*f1^2*f3^-1*f1^-1,
f3*f2^2*f3^-1*f2^-1, f3^2*f1*f3^-1*f1^-1, f3^2*f2*f3^-1*f2^-1]] .

Acknowledgements

I wish to express my gratitude to all those who contibuted directly or indirectly towards this research work. First and foremost, my sincere thank goes to the Almighty God for his grace and mercies upon me, and for directions throughout this work. Many thanks to Professor F. I. Njoku of the University of Nigeria Nsukka, and Professor Jamshid Moori of North-West University, South Africa for introducing me to Group theory. I am grateful to the African Institute for Mathematical Sciences (AIMS) South Africa for exposing me to Postgraduate courses in Mathematics, and the programming skills instilled in me.

I am deeply indebted to the University of Warwick, and its Institute of Mathematics represented by Professor Colin Sparrow for providing me bursary for this Masters programme. My profound gratitude goes to my Supervisor, Professor Derek Holt for his top-notch supervision. Many thanks to my Adviser Dr. Saul Schleimer for his constant support. I am grateful to J. H. C. Whitehead, J. Nielsen, E. S. Rapaport, P. J. Higgins, R. C. Lyndon, J. McCool and many other Mathematicians that contributed in one way or the other towards making Automorphisms of Free Groups worth investigating.

A million thanks to my father Mazi Christopher Anabanti, siblings, Uncle and Aunty (Drs. Okoro and Catherine Chima-Okereke) for their love and encouragement. Special thanks to my cousins Chinyere, Chibisi, Onyekachi, Uchechi and Iheomachi; and my friends Ghaleo, Bibekananda, Alex and David...to mention but a few. A shout-out to all my friends and relatives unmentioned. Finally, I dedicate all these works to my late mother Mrs Elizabeth Anabanti.

References

[AFV08]   Heather Armstrong, Bradley Forrest, and Karen Vogtmann, *A presentation for Aut(F_n)*, Journal of Algebra **11** (2008), 267–276.

[Bog00]   O. Bogopolski, *Classification of automorphisms of the free group of rank 2 by ranks of fixed-point subgroups*, Journal of Group Theory **3** (2000), 339–351.

[GAP13]   The GAP Group, *GAP – Groups, Algorithms, and Programming, Version 4.6.5*, 2013.

[Ger84]   S. M. Gersten, *A presentation for the special automorphism group of a free group*, Journal of Pure and Applied Algebra **33** (1984), 269–279.

[HL74]   P. J. Higgins and R. C. Lyndon, *Equivalence of elements under automorphisms of a free group*, Journal of London Mathematical Society **8** (1974), 254–258.

[HMM05]   R. M. Haralick, A. D. Miasnikov, and A. G. Myasnikov, *Heuristics for the Whitehead minimization problem*, Experimental Mathematics **14** (2005), 7–14.

[Joh97]   D. L. Johnson (ed.), *Presentations of Groups*, London Mathematical Society Student Texts, no. 15, Cambridge University Press, The Edinburgh Building, Cambridge, CB2 2RU, United Kingdom, 1997.

[LS77]   Roger C. Lyndon and Paul E. Schupp (eds.), *Combinatorial Group Theory*, A Series of Modern Surveys in Mathematics, no. 89, Springer-Verlag, Berlin Heidelberg New York, 1977.

[McC74]   J. McCool, *A Presentation for the automorphism group of a free group of finite rank*, Journal of London Mathematical Society **8** (1974), 259–266.

[Mes74]   Stephen Meskin, *Periodic automorphisms of the two-generator free group*, Proceedings of the Second International Conference on the Theory of Groups **372** (1974), 494–498.

[MKS76]   W. Magnus, A. Karass, and D. Solitar (eds.), *Combinatorial Group Theory: Presentations of Groups in terms of generators and relations*, Dover Publications, Inc., New York, 1976.

[MM04]   Alexei D. Miasnikov and Alexei G. Myasnikov, *Whitehead method and generic algorithms*, Contemporary Mathematics **349** (2004), 89–114.

[Neu32]   B. H. Neumann, *Die automorphismengruppe der freien Gruppen*, Math. Ann. **107** (1932), 367–386.

[Nie24]   J. Nielsen, *Die Isomorphismengruppe der freien Gruppen*, Math. Ann. **91** (1924), 169–209.

[Rap58]   E. S. Rapaport, *On free groups and their automorphisms*, Acta Math. **99** (1958), 139–163.

[Whi36a] J. H. C. Whitehead, *On certain sets of elements in a free group*, Proceedings of London Mathematical Society **41** (1936), 48–56.

[Whi36b] ———, *On equivalent sets of elements in a Free Group*, Annals of Mathematics **37** (1936), 782–800.